solar energy and building

solar energy and building

s v szokolay

Second Edition

The Architectural Press, London
Halsted Press Division
John Wiley & Sons, New York

First published 1975 in Great Britain by
The Architectural Press Ltd

Reprinted (*twice*) 1976
Second edition 1977
© S. V. Szokolay 1975

ISBN 0 85139 570 8

Published in the U.S.A. by
Halsted Press, a Division of
John Wiley & Sons, Inc.,
New York.

Library of Congress Cataloging in Publication Data

Szokolay, S V

    Solar energy and building. 2nd edition

    Includes bibliographical references and index.

    1. Solar heating.   2. Solar energy.   3. Solar
houses. I. Title.
TH7413.S96   697'.78   77–21700
ISBN 0-470-99235-2 (Wiley)

Filmset and printed in Great Britain by
BAS Printers Limited, Over Wallop, Hampshire

Contents

Preface
Preface to second edition
Glossary

Part 1 Context and Principles

Part 2 Collective Methods

Part 3 Uses

Part 4 Sun and Building

Part 5   Solar Houses

Part 6   Planning Implications

Part 7   Economics and Prospects

Part 8   Solar Heat Industry

Part 9   Design Guide

Part 10   Progress in Some Countries

Part 11   Development in Applications

Part 12   Theory and Methods

Preface

This book is the by-product of some four years of research work carried out at the Polytechnic of Central London. Early in 1974 we felt that the piles of information collected should be made available to a broad readership, particularly as much of this information is in conference papers or documents with limited circulation, or indeed in publications now out of print. In preparing the text I tried to keep three aims in mind:

to collect and present the material scattered (and often hidden) in many sources

to summarise the product of a very large number of research workers in a systematic form

to give a conceptual understanding of the problems and the solutions and thus, by describing what has been done, help to generate new ideas.

The importance and the potential of solar energy utilisation cannot be over-emphasised. At a meeting of the rather conservative Institute of Fuel in September 1974 in Brisbane, it was suggested that by the year 2000, 30% of our energy needs should be supplied from solar sources, 30% by atomic power, and that the conventional methods would only contribute the remaining 40%.

This is quite a tall order. It won't happen, unless it is made to happen. The more people work on it, and the greater the range of ideas and solutions, the greater the likelihood of success.

Much work has been done on the subject. And, as the old Hungarian saying has it, 'Man learns by his own mistakes, but the wise one by the mistakes of others'. The reader is encouraged to take it further and do better than the solutions here described.

My acknowledgements are due to so many that I won't even start. It must however be acknowledged that very little of what I have written is my own contribution. I have only put it together. And my sincere apologies are due to all those workers to whom I could not do justice: long years of research are often mentioned in one brief sentence. I don't claim to be comprehensive. Much has been omitted by choice, as too specific or narrow or as outside the scope stated by the title, but there must also be quite a lot of work of which I am not aware. My only possible excuse is that this is a new field, we are breaking new ground. One day, in the not too distant future, it will become a discipline, a defined body of knowledge. Today we can still enjoy the freedom of pioneers.

Steven V Szokolay
Brisbane, October 1974

Preface to the second edition

There is no exaggeration in the statement that more has been done in the field of solar energy research and development since June 1974 (when the manuscript of this work was first completed) than the total work up to that date.

I am grateful to the publisher for giving me the opportunity to update this work. A few minor corrections have been made and the new added parts are an attempt to outline the developments of the last two and a half years.

Part 10 gives a country-by-country survey of developments, whilst Part 11 is organized according to particular areas of application. Part 12 gives a review of developments in theoretical subjects and in computation.

Steven V. Szokolay
Brisbane, January 1977

Glossary

of a few not generally known terms

a/e ratio
the ratio of absorption coefficient for solar radiation to the emission coefficient at operating temperature, a measure of the selectivity of absorber surfaces.

coefficient of performance
used in relation to heat pumps or refrigeration machines, it is the ratio of the amount of heat delivered to the energy input to the machine. Its value is normally between 3 and 6 for compression refrigerators and less than 1 for absorption machines.

collection efficiency
a ratio of the amount of useable energy supplied by the system to the amount of solar energy incident on the collector surface over a period stated (instantaneous, daily, annual).

degree-days
a climatic index: the cumulative temperature deficit for a given period (year or month), given in relation to a specified reference temperature (cf section 7.2).

figure of merit
a comparative index figure, obtained as a quotient: the value of energy actually saved by an installation over ten years, divided by the cost of that installation. A FM of 1 is often taken as the criterion of feasibility.

low grade energy
a relative term: radiant energy of a longer wavelength (emitted by a body of lower temperature) is of a *lower grade* than an identical quantity of a shorter wavelength (emitted by a hotter body). In relation to heat it means a lower temperature.

present worth
A sum *X*, expressed in monetary terms, to be paid in the future, will be worth less than its present monetary value. Invested now, it would earn interest; thus to find the true value of *X* in today's terms one would calculate it on the basis of present worth + interest. This is done by using the compound interest formula in reverse.

selective absorber
a dark surface having a high absorption coefficient for short wave solar radiation, but a low emission coefficient for long wave low temperature radiation (cf section 2.4).

specific heat loss rate
an index of the thermal properties of a building, given as the total (conduction + convection) heat loss rate per unit temperature difference (W/degC).

thermosyphon
the hydraulic system in which a fluid circulation is caused by temperature (thus density-) differences in the fluid.

(others of only immediate significance are explained in the text as they occur)

Symbols used:

	defined or introduced in section
α (alpha) = solar azimuth angle	1.9, 4.3
β (beta) = angle of incidence	1.9
γ (gamma) = solar altitude angle	1.9, 4.3
Δ (delta, cap) = as prefix: difference or change in some quantity	2.3
δ (delta) = horizontal shadow angle	4.3
ε (epsilon) = vertical shadow angle	4.3
η (eta) = efficiency	2.5, 2.8
θ (theta) = transmission coefficient	2.3
λ (lambda) = wavelength	1.6
σ (sigma) = Stefan-Bolzmann constant	2.11
Σ (sigma, cap) = 'sum of' the expression following	1.4
τ (tau) = solar gain factor	4.1
ϕ (phi) = geographical latitude	1.9
ψ (psi) = tilt of plane from horizontal	1.9
ω (omega) = orientation (in azimuth angle)	4.3
a = absorption coefficient	2.3, 2.4, 4.1
c = thermal conductance (W/m² degC)	2.3
e = emission coefficient	2.3, 2.4, 4.1
f = film, or surface conductance (W/m² degC)	2.3, 4.2
h = number of hours	2.3
s = specific heat, volumetric (Wh/lit. degC)	2.4
t = temperature (°C)	—
Δt = temperature difference (degC)	2.3, 2.4
t_e = sol-air excess temperature	4.1
t_f = flow temperature	2.4
t_i = indoor air temperature	4.1
t_o = outdoor air temperature	2.4
t_r = return temperature	2.4
t_s = sol-air temperature	4.1
A = area (m²)	2.3
C = heat capacity (Wh/degC)	2.3
Cr = concentration rate	2.9
F = flow rate (litre/h)	2.4
FM = figure of merit	7.1
H = heat energy (Wh)	1.4
H_g = heat gain	2.3
H_i = heat input	2.3
H_l = heat loss	2.3
I = intensity or density of energy flow rate (W/m²)	2.3
I_{hf} = horizontal diffuse intensity	1.9
I_{hr} = horizontal direct intensity	1.9
I_{ht} = horizontal total intensity	1.9
I_{nr} = direct intensity on a plane normal to the direction of flow	1.9
I_{pf} = diffuse intensity on a given plane	1.9
I_{pr} = direct intensity on a given plane	1.9
I_{pt} = total intensity on a given plane	1.9
Q = quantity of heat flow rate (W)	1.4
Q_i = heat input rate	2.3
Q_l = heat loss rate	2.3
Q_s = solar heat gain rate	4.1
T = temperature, absolute (°K)	2.8
U = thermal transmittance (W/m² degC)	2.3

Part 1　Context and Principles

1.1
Energy consumption

Perhaps the most important single characteristic of our society, at least from an objective, material point of view, is that it is based on an abundant and ever increasing supply of energy. Without this our industry, transport, even agriculture or urban domestic life could not exist.

For thousands of years society existed on human and animal labour. Early inanimate sources of energy, such as windmills and water-wheels gave a significant increase in working rate (or power) but the qualitative jump came about only in the 17th and 18th century. It may be illustrative to compare the magnitude of a few power sources: [1]* (For definition of W ie *watt* see section 1.2.)

man	80 W	(up to 300 W for short periods)
ass	180 W	
mule	370 W	
ox	500 W	
horse	750 W	
watermill	1·5–3·8 kW	(5 m dia. overshot wheel)
windmill	1·5–6·0 kW	(typical post windmill)
steam engine	5·2–7·5 kW	(early stationary type)
1000 cc car	45–60 kW	
steam turbine	up to 100 MW	

Development of the internal combustion engine and of the various turbines increased both the power of the individual units and the total number of units, thus their aggregate output and their corresponding fuel consumption. The exponential growth of fuel consumption started with the industrial revolution in the 18th century and was unchecked until quite recently. Fig. 1.1 shows the growth of the world's energy consumption for the 40-year period 1931–1971.

Today over 98% of our energy comes from fossil fuels: coal, oil and natural gas. However large, the stock of fossil fuels is finite. Their continued consumption means living on our capital. The present rate of exploitation is clearly untenable. Both coal and oil are not only fuels, but also important raw materials of our chemical industries. Their use for fuel is rather shortsighted, to say the least.

There are several ways to produce a 'stocktaking' balance sheet. The simplest of these is the *static index*. This is produced by dividing the present annual consumption into the amount of known reserves. Although the result is in a number of years, it is not a forecast, just an index. [2] This way we get 2300 years for coal and 31 years for oil.

* Figures in square brackets, eg [1], refer to references given at the end of each part.

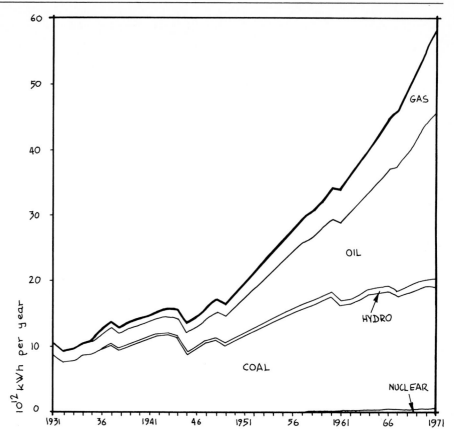

1.1
world energy production, 1931–71. Based on *UN Statistical Yearbook*, 1951, 1966 and 1972 (1951–54 data by interpolation only

The *exponential index* is produced in a similar way, but taking into account the exponential growth of consumption. This gives 111 years for coal and 20 years for oil.

Actual forecasts take into account estimates of future discoveries of stocks as well as a reducing rate of consumption, dictated by price increases expected with growing scarcity. Fig. 1.2 shows 'high' and 'low' estimates for future oil and coal production. [3]

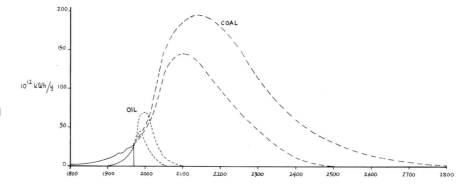

1.2
world coal and oil production cycles showing 'high' and 'low' forecasts. Based on *Resources and Man*, US National Academy of Sciences and National Research Council, 1969). Conversions used: 1 tonne coal = 8000 kWh; 1 barrel of oil = 1870 kWh

Until recently the availability of energy 'on tap' was taken for granted. The average man was not even aware of the intricate network of fuel production and industry serving his comfort. The division of labour, carried to an extreme, bred the short-sighted attitude of 'never mind where it comes from, as long as I get it', which prevails in our consumer society. We no longer realise the inherent value of things we possess or consume. I suggest that this is at least one of the causes of 'alienation', the divorce of one's personal life from society as a whole and from the natural processes on which we all depend. This breeds an almost parasitically egotistic attitude. As a result of this we tend to be preoccupied with and we even create superficial pseudo-problems. We lead a life in spite of, rather than in sympathy with the ecological system of our thin bio-sphere.

Our most basic energy requirement is food. The daily per capita food consumption varies between wide limits:

UK average consumption	3·30 kWh/day	(1 kilo-calorie = 1·163 Wh)
FAO recommended minimum	2·60	
our 'minimum'	1·75	
average in India	0·90	

The world total annual food production is approximately 2200×10^9 kWh, which would give about 1·7 kWh/day to all $3·5 \times 10^9$ people of the earth, but the distribution is very uneven. In fact some 80% of the world's population is undernourished.

The annual total food consumption in the UK is about 67×10^9 kWh. This is less than 4% of the total energy consumption, which is 1670×10^9 kWh/year. According to final uses, it is distributed as follows: [4]

industry	705×10^9 kWh/year
domestic	415
transport	341
public services	100
miscellaneous	86
agriculture	23
total in final use	1670×10^9 kWh/year

This huge quantity comes from the following sources:

coal	1145×10^9 kWh/year
oil	1200
natural gas	210
nuclear electricity	70
hydro-electricity	25
total produced	2650×10^9 kWh/year

The difference between use and production represents losses in conversion and distribution.

The domestic use, ie the energy used for water and space heating, is about 25% of the total. It is this quantity which can be strongly influenced by buildings and building services, thus it is the direct responsibility of architects and the building professions.

1.2 Physical quantities

Before going further it may be useful to revise some of the relevant physical principles.

Energy is the potential for carrying out work and is measured in the same units as work. In the System International this unit is derived from the three basic assumed and agreed units: length, mass and time. The logical way of this derivation is useful also to show the relationship of the physical quantities themselves.

length	**m** (metre)	
mass	**kg** (kilogramme)	
time	**s** (second)	
velocity	the length of movement in unit time	**m/s**
acceleration	change of velocity in unit time	**m/s²**
momentum	the state of a body, the product of its mass and its velocity	**kg m/s**
force	measured by its effect: the change in momentum per unit time, or the acceleration given to unit mass:	**kg m/s²**
	This unit is known as the Newton:	**N**
work, energy	measured as the product of a force and the distance over which it has acted: $N \times m$:	**kg m²/s²**
	This unit is known as the Joule:	**J**
power	measured as the rate at which work is done or the rate of energy flow:	**kg m²/s³**
	or J/s, known as Watt:	**W**

When the rate of energy flow of 1 W is maintained for an hour (3600 seconds), the amount of energy spent is 1 Wh (watt-hour). Thus Wh is an energy unit, of the same physical dimension as J:

1 Wh = 3600 J

Although the official SI unit of energy or work is the Joule, we adopt the watt-hour (Wh) as a much more practical unit. Its multiple, the kWh is in general use as a 'unit' of electricity. Even the non-scientific mind can get a 'feel' of its magnitude:

—a 1 kW electric bar heater, used for one hour, consumes 1 kWh electricity and emits 1 kWh of heat

—a 100 W incandescent electric lamp used for 10 hours consumes 1 kWh electricity and emits some 950 W of heat and 50 W of light.

Various sources use a multitude of different energy and power units. It is suggested that all such data should be converted to Wh and W respectively, in order to achieve comparability. The most important and often encountered units can be converted using the following factors: [5]

Energy or work

1 J	=	0·000 278 Wh
1 kJ	=	0·278 Wh
1 erg	=	$0·278 \times 10^{-10}$ Wh
1 ft lbf	=	0·000 377 Wh
1 cal	=	0·001 163 Wh
1 m kgf	=	0·002 726 Wh
1 Btu	=	0·293 Wh
1 kcal	=	1·163 Wh
1 hp h	=	0·746 kWh
1 therm	=	29·33 kWh

Power or rate of energy flow

1 erg/s	= 0·000 000 1 W
1 Btu/h	= 0·293 W
1 kcal/h	= 1·163 W
1 ft lbf/s	= 1·355 82 W
1 cal/s	= 4·186 8 W
1 hp (metric)	= 735 W
1 hp	= 746 W
1 therm/h	= 29 307 W

1.3
Forms of energy

Potential energy (or positional energy) of a body in a position from which it can freely fall is the same as the work required to lift it to the same position against the gravitational force. It is the product of the gravitational force acting on the body (its 'weight') and the available height:

mass × gravitational acceleration × height = energy
  kg ×   m/s²   × m = J

In a mechanical system the sum of kinetic energy and potential energy is constant. When a bullet leaves the gun (aimed upwards) it has a maximum of kinetic energy and no potential energy. When it reaches the peak of its ballistic path, its kinetic energy is reduced but it has acquired a potential energy. In free fall this will be converted again into kinetic energy.

Sound energy is a special case of mechanical energy, transmitted as vibration of particles in a substance, each particle rapidly converting kinetic to potential energy and back. The magnitude of energy present in sound is minute. Conversational speech would mean the emission of about 10^{-5} W and a pneumatic rivetter about 1 W, ie 1 J per second. A sound level of 120 dB would correspond to the flow of 1 J of energy through a 1 m² area in one second, ie 1 W/m².

Heat is a form of energy contained in bodies as molecular motion and causing the body to show a certain temperature. With the increase of this molecular motion the solid will liquefy and with still further increase it may evaporate. Specific heat of a material is the amount of heat energy required to increase the temperature of unit mass by one degC. It is measured in Wh/kg degC (or J/kg degC).

Chemical energy of compounds is the same as the energy required to produce these compounds from their basic components. Processes requiring energy (heat) to build up such compounds are termed endothermic processes. Exothermic processes are those which release energy (heat). Fuel materials are compounds of high energy content which can be readily released by combustion. Combustion or oxidation are exothermic processes. The chemical energy that can be released as heat is measured as the calorific value, in units of Wh/kg or Wh/m³ (J/kg or J/m³).

Electrical energy. If a power of 1 W is operated for 1 s, unit work (1 J) is done or unit energy has been expended

$$J = W \times s$$

Electrical energy is most frequently produced by the application of mechanical energy (in generators or dynamos) or by the release of chemical energy (in batteries). It can be converted into almost any other forms of energy.

Radiant energy is a term applied to all forms of electro-magnetic radiation from low frequency (long wave) electric or radio waves, through heat, light and X-rays, to various gamma and cosmic radiations of very high frequency and short waves (see Fig. 1.4).

1.4
Principles of thermodynamics

In simple terms the first law of thermodynamics is the principle of the conservation of energy. Energy cannot be created or destroyed, only converted from one form to another (except in nuclear processes, where small parts of the matter used may be converted to energy).

The second law of thermodynamics (formulated by Clausius in 1850) states that heat (or energy) transfer can take place spontaneously in one direction only: from a hotter to a cooler body. Kelvin extended the law to state that it is impossible to construct a device which would operate in a cycle and perform work (converting heat to mechanical work) with a single reservoir. Both a sink and a source are necessary. In other words: heat must flow from a source to a sink and only a part of this flow can be converted to work.

This law governs the operation of expansion (Stirling) engines, steam or internal combustion engines and sets a theoretical limit to their efficiency. It will be referred to again in conjunction with heat pumps (section 2.8).

Entropy is a rather difficult concept to understand. It is the absence of energy. A high temperature means small entropy. The absolute zero temperature would mean an infinitely large entropy. In all physical processes the entropy is positive (or possibly zero). Energy is always degraded. Only biological systems are said to show a negative entropy, ie an increase in energy content, a growing order. Entropy is randomisation.

The reciprocal of the absolute temperature, $\frac{1}{T}$ is an index of randomisation. With a change in heat content of dQ at a temperature T, the entropy change is $\frac{dQ}{T}$. The net entropy change in any physical process is positive or zero $\Sigma \frac{dQ}{T} \geqslant 0$ [1] (the capital *sigma*: Σ is generally used to denote 'sum of').

1.5
Global energy flow

Earth receives radiant energy from the sun at the rate of 173×10^{15} W.* (For comparison: taking the annual energy consumption of humanity as 61×10^{15} Wh and dividing this by the number of hours in a year, $24 \times 365 = 8760$, we get a rate of consumption of 7×10^{12} W.)

Earth also emits an identical amount. This is a condition of equilibrium. The emission depends on the earth's temperature. The temperature at which the emission equals the input is the equilibrium temperature, ie the temperature of earth as we know it. Should the input change for some reason, the equilibrium temperature would also change.

* Projected area of earth = $(6.3 \times 10^{6})^{2} \times 3.14 = 124 \times 10^{12}$ m²
Solar constant = 1395 W/m² thus
$124 \times 10^{12} \times 1395 = 173 \times 10^{15}$ W

1.3
flow of energy through the terrestrial system

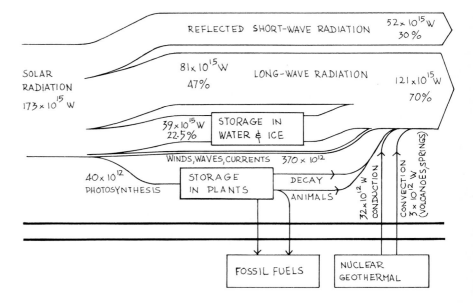

Some 30% of the incoming radiation is reflected without change in wavelength. About 47% is absorbed by the atmosphere and the earth's surface, causes a temperature increase and is subsequently re-radiated to space. Only the remaining 23% enters the terrestrial system and becomes the motive force of winds, currents, waves, shapes our climate and causes the hydrological cycle. Ultimately this will also be re-radiated to space.

A mere 0·02% of the total, or 40×10^{12} W, enters the biological system through photo-synthesis in plants and other 'producer' organisms (cf section 2.1 and Fig. 2.9).

Fig. 1.3 illustrates this global flow-pattern. [6]

A small proportion of the energy stored as chemical energy in plant and animal body tissue has over millions of years accumulated under favourable geological conditions as coal and mineral oil, forming our stock of fossil fuels. This is 'capital' stock. The rate of formation of fossil fuels (if any at all) is negligible compared with the rate of consumption. If we want to avoid further depletion of our fossil fuel stocks, what we ought to do is to tap the flow of these huge quantities of energy and re-route some of it to do work for us, before being dissipated and re-radiated to space. (Fig. 1.3/a)

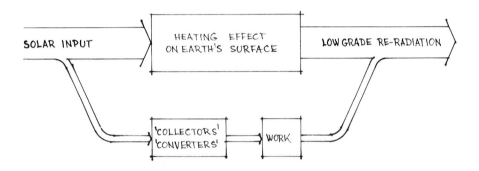

1.3/a
re-routing part of the energy flow

1.6
Solar radiation: quality

The emission spectrum of black body radiators is determined by their temperature. The spectrum of solar radiation outside the atmosphere very nearly corresponds to the emission of a black body at 6000 °K. It is an almost continuous spectrum from about 200 nm (nano-metre = 10^{-9} m) ultra-violet, to 3000 nm infra-red, with a strong peak around 500 nm. Atmospheric absorption is to some extent selective, changing not only the quantity, but also the spectral composition of the radiation received. Fig. 1.4 shows the two spectra in relation to the whole of the electro-magnetic radiation spectrum. The gaps in the lower curve show the characteristic absorption bands of our atmospheric gases: oxygen, nitrogen, carbon dioxide, but mainly of water vapour.

Electromagnetic radiation (such as light) shows dual characteristics, which are explicable in terms of the corpuscular and wave theories. The energy content of radiation is determined by its wavelength.

The shorter wavelengths represent a higher grade energy. It will be shown later that all of the solar radiation can be considered for conversion to heat, but only the short wave, high energy components will be able to produce a photoelectric effect.

1.7
Solar radiation: quantity

The intensity of radiation reaching the upper limits of our atmosphere shows some variations, but the mean value, 1395 W/m² is taken as the *solar constant*. Sun spot activity may change the energy output of the sun itself by $\pm 2\%$ and there is a variation of $\pm 3·5\%$ due to the variation of the earth—sun distance (152×10^6 km at aphelion and 147×10^6 km at perihelion).

Atmospheric absorption reduces this intensity to an extent depending partly on the length of travel through the atmosphere and partly on the state of the air mass (cloudiness, suspended particles). When the sun is at a low altitude angle, the intensity is less. With a zenith position the intensity measured on a horizontal plane may approach 1 kW/m² (at sea level).

Fig. 1.5 shows the intensity of solar radiation on a horizontal plane as a function of solar altitude angle, at sea level and at various heights. [7]

The annual total amount of radiation received at a given location depends on its geographical latitude and on local climatic factors. Fig. 1.6, the solar radiation map of the earth, gives a rough indication of what can be expected at various locations.

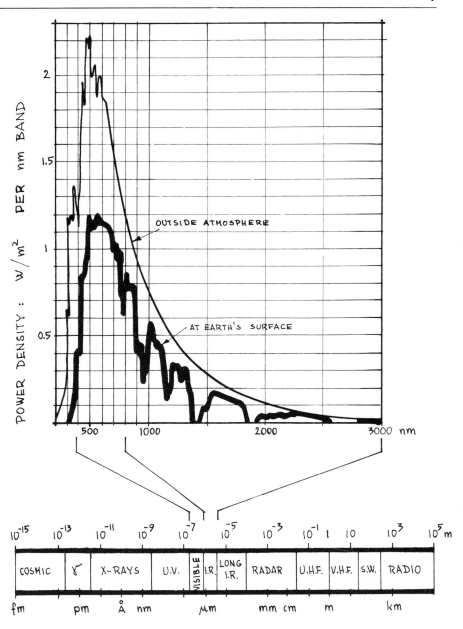

OUTSIDE ATMOSPHERE

AT EARTH'S SURFACE

POWER DENSITY : W/m² PER nm BAND

500 1000 2000 3000 nm

10^{-15}	10^{-13}	10^{-11}	10^{-9}	10^{-7}	10^{-5}	10^{-3}	10^{-1}	1	10	10^{3}	10^{5} m

COSMIC	γ	X-RAYS	U.V.	VISIBLE	I.R.	LONG I.R.	RADAR	U.H.F.	V.H.F.	S.W.	RADIO

fm   pm   Å   nm   μm   mm cm   m   km

1.4
solar radiation spectrum related to the full electro-magnetic spectrum

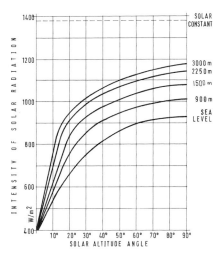

1400   SOLAR CONSTANT
3000 m
2250 m
1500 m
900 m
SEA LEVEL

INTENSITY OF SOLAR RADIATION
W/m²

10° 20° 30° 40° 50° 60° 70° 80° 90°
SOLAR ALTITUDE ANGLE

1.5
variation of direct solar intensity with height and angle of incidence (original by IS Groundwater, in Btu/ft²h and ft units)

1.6
solar radiation map of the world: average annual total radiation in kWh/m² year [11]

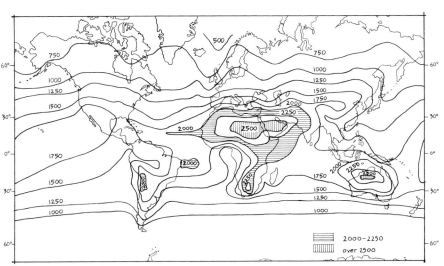

2000-2250
over 2500

There is however a variation in the annual distribution of solar radiation at any particular point. Generally, in equatorial locations the annual pattern is fairly even, whereas at higher latitudes there is a pronounced summer peak, as shown by Fig. 1.7. [8]

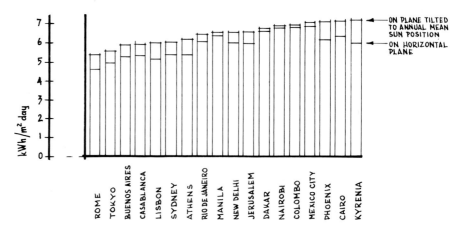

DAILY AVERAGE SOLAR RADIATION

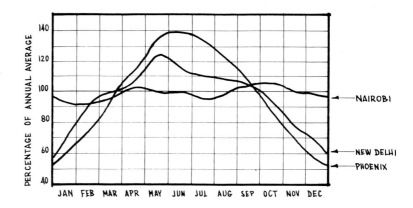

1.7
daily average solar radiation at different locations and annual variation of solar radiation (after Hobson in *World Symposium on Applied Solar Energy*)

1.8
Solar radiation: data

Meteorological records may give the hourly totals or hourly average intensities (Wh/m²h or W/m²) measured on a horizontal plane. Some stations record the total radiation and the diffuse component separately.

Stations actually measuring and publishing hourly data are rather few and far between. Often one can only find daily totals or possibly one average day's total for each month, or the total cumulative amount received in each month. The annual total is often used to characterise a particular climate. An excellent summary of world radiation data is given by Löf. [13]

If there are no radiation data available, but the duration of sunshine is recorded, one can estimate the daily total radiation, using the expression given by Glover and McCulloch: [9]

$$Q = Q_{sc} \left(0.29^* \times \cos \phi + 0.52^* \frac{n}{N}\right)$$

where Q = daily total radiation on a horizontal plane (Wh/m² day)
Q_{sc} = 'solar constant' per day
ϕ = geographical latitude
N = possible⎱
n = actual ⎰ sunshine hours per day

the value of Q_{sc} can be taken as 9830 Wh/m² day thus eg for latitude 52° (London):

* Ometto suggests the use of constants 0.26 and 0.51 respectively [10].

$$Q = 9830 \left(0.29 \times \cos 52 + 0.52 \frac{n}{N}\right)$$

$$Q = 1755 + 5111 \frac{n}{N}$$

1.9
Incident radiation

If the radiation intensity incident on a tilted plane is to be calculated, the total radiation as measured on a horizontal plane must be divided into its direct and diffuse components. The direct component will then be handled vectorially. The diffuse radiation incident on the tilted plane will be proportionate to the fraction of the sky hemisphere to which the plane is exposed.

The calculation method is as follows:

records give horizontal total intensity*: I_{ht}
and horizontal diffuse intensity: I_{hf}

The horizontal direct intensity is the difference of the two

$$I_{hr} = I_{ht} - I_{hf}$$

The position of the sun at the date and time considered is determined in terms of an altitude angle (γ) and an azimuth angle (α), the latter being measured from the North. These angles can be found from various almanacs, may be estimated from various solar charts or sun-path diagrams (see section 4.3), or can be calculated from the two astronomical equations:

$$\sin \gamma = \sin d \times \sin \phi - \cos d \times \cos \phi \times \cos t$$
$$\sin \alpha \times \cos \gamma = \cos d \times \sin t$$

where d = declination (varying from 0° at the equinoxes, to +23·5° on June 21 and −23·5° on Dec 21. See Fig. 1.8)
  ϕ = geographical latitude
  t = hour angle (15° for each hour)†

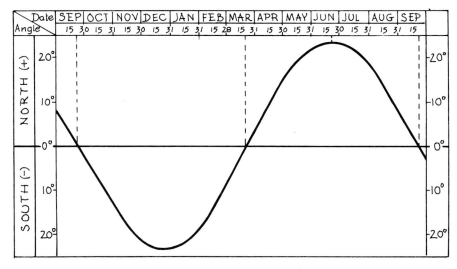

1.8
the sun's declination ie the angle between the plane of the earth's orbit and the equatorial plane [12]

First the direct intensity on a plane normal to the direction of radiation is found

$$I_{nr} = \frac{I_{hr}}{\cos (90 - \gamma)} = \frac{I_{hr}}{\sin \gamma}$$

The angle of incidence (β) on the particular tilted plane can be found from

$$\cos \beta = \sin \gamma \times \cos \psi + \cos (\omega - \alpha) \times \cos \gamma \times \sin \psi$$

where ω = orientation of plane (from North)
  ψ = tilt of plane from horizontal

The direct intensity on the plane is

$$I_{pr} = I_{nr} \times \cos \beta$$

The diffuse intensity on the plane is

$$I_{pf} = I_{hf} \frac{1 + \cos \psi}{2}$$

Total incident intensity on the plane is the sum of the two

$$I_{pt} = I_{pr} + I_{pf}$$

* Most instruments measure the total radiation. The direct component can be eliminated by simple shading, and the remainder is measured as the diffuse radiation.
† Ignoring the slight variation due to changes in orbital speed, expressed by the 'equation of time'.

10

References

1 Ubbelohde, A R
Man and energy
Penguin, 1963

2 Meadows, D H and Meadows, L M
The limits to growth
Earth Island, 1972

3 Nat. Academy of Sciences and Nat. Research Council
Resources and man
Freeman & Co. 1969

4 Dept. of Trade and Industry
UK energy statistics
HMSO, 1973

5 National Physical Laboratory
Changing to the metric system
HMSO, 1967

6 Scientific American
Energy and power (special issue)
September, 1971

7 Groundwater, I S
Solar radiation in air conditioning
Crosby Lockwood, 1957

8 Hobson, J E
The economics of solar energy
in Proc. World Symposium on applied solar energy
Uni. of Arizona, Stanford Research Inst. 1955

9 Glover, J and McCulloch, J S G
The empirical relation between solar radiation and hours of sunshine
Q.J. Royal Meteorological Soc. **84** (1958)

10 Ometto, J C
Etude des relations entre le rayonnement solaire global, le rayonnement net et l'ensoleillement
paper E 21
'The sun in the service of mankind' Conf. Paris, 1973

11 Ransom, W H
Solar radiation and thermal effects on bldg materials
BRS Tropical Building Studies No. 3
HMSO, 1962

12 Phillips, R O
Sunshine and shade in Australasia
Com'wealth Exp. Bldg. Stn.
Technical Study No. 23
Sydney, 1951.

13 Löf, G O G
World distribution of solar radiation
Uni. of Wisconsin, 1966

Part 2 Collective Methods

2.1
Chemical conversion

In this chapter the various methods available for the conversion of radiant energy into a useable form will be examined.

The most important forms of chemical conversion of solar energy are the photo-biochemical processes. Biological organisms are classified by ecologists as 'producers' and 'consumers'. Producers, ie plants and algae synthesise carbohydrates from carbon dioxide and water, absorbing some of the incident solar energy and storing it in chemical bonds. Consumers (herbivorous animals) use the body tissue of producers as food. (Carnivorous animals can thus be taken as secondary consumers, as they eat animals which are themselves consumers.) The materials of this food chain are recycled, as shown by Fig. 2.1, but the energy flows through in one direction. [1]

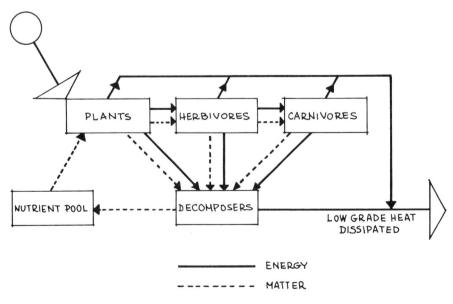

2.1
a typical food chain showing the flow-through of energy and the recycling of matter [1]

The generalised schema of photosynthesis is

$$\text{carbon dioxide} + \text{water} \; \frac{\text{sunlight}}{\text{chloroplast}} = \text{carbohydrate} + \text{oxygen}$$

The chloroplast is acting as a catalyst only.

One specific example:

$$6(CO_2 + H_2O) \; \frac{5 \text{ kWh/kg solar radiation}}{\text{chlorophyl}} = C_6H_{12}O_6 + 6O_2$$

When this material passes through the food chain, there is a significant loss at every step, as animals use some of its energy content for locomotion and discard much of it as excreta. The following tabulation shows a typical food-chain in a freshwater lake: [2]

	algae	zooplankton	fish	carnivora	man
body matter	1 kg	150 g	30 g	6 g	1·2 g
energy content	5 kWh	750 Wh	150 Wh	30 Wh	6 Wh

Producers, on the global scale, store solar energy (as section 1.5 above) at the rate of 40×10^9 kW. This corresponds to an annual quantity of energy of

$$40 \times 10^9 \times 24 \times 365 = 350 \times 10^{12} \text{ kWh}$$

Of this about $2·2 \times 10^{12}$ kWh is used as food material by the human population (some 0·63%).

There are a number of other biochemical processes which could be developed. Hydrogen, with its free valency, is a high-energy substance, it releases a large amount of heat when combining with oxygen. Algae and other green plants can produce hydrogen, in the presence of the appropriate enzymes, using some of the incident solar energy and using water as the reagent.

Non-biological catalytic decomposition of water is also possible, producing hydrogen and oxygen in separate compartments.

Certain dye materials undergo a reversible reaction, taking up some energy when exposed to sunlight and releasing it when in the dark. One example of these is the blue dye thionine, in the presence of ferrous ions in water. The light reaction oxidises the ferrous ions to ferric ions, which react with the dye, reducing it to a colourless substance. The process is reversed in dark, when an electric potential difference is developed.

The explanation for photochemical reactions, in general terms, is that when a quantum of radiant energy is absorbed by an atom, it becomes 'excited'. This is an unstable, transitional state. The surplus energy may be re-emitted as light, but it may also cause chemical changes of one of three kinds:

1   photo-dissociation (eg hydrogen from water)
2   photo-combination (eg dyes or biochemical synthesis)
3   photo-sensitisation, where the excited molecule acts as an energy carrier and will induce chemical changes in other molecules

2.2 Electrical conversion

Solar radiation may be converted directly into electricity using two possible processes:

1   thermoelectric conversion
2   photoelectric conversion

Each of these has several forms.

1/a Thermionic generators

When an electrode is heated, some of its electrons will acquire enough energy to escape. It becomes an electron-emitter, a *cathode*. Another electrode placed close to the cathode, if sufficiently cooled, will readily receive the emitted electrons, thus it becomes an *anode*. If the anode is connected to the cathode through a circuit containing an external load, a current will flow and work can be produced. (Fig. 2.2)

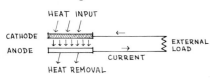

2.2
thermionic generator

The electrode material may be tungsten or caesium-coated silver oxide. The electrodes are kept at a distance of a fraction of a mm, either in vacuo or in caesium vapour.

Significant current can only be produced with very high cathode temperatures (1000–2500°C), thus thermionic generators can only be used in conjunction with concentrating devices with concentration ratios of 500 to 3000.

Under bright sunshine conditions the thermionic cell potential would be in the order of 1 V only, but many cells can be connected in series to give a higher voltage, in D/C. Converters can be used to produce A/c which then can be transformed to any desired voltage.

Theoretically the limit of efficiency is around 30%. In practice efficiencies of 6 or 8% have been obtained. The cooling medium can also serve a useful purpose, eg the production of hot water. [3]

1/b
Thermocouples and thermopiles

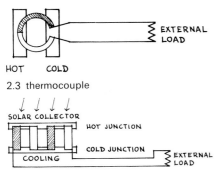

2.3 thermocouple

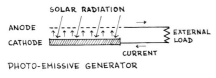

2.4 solar thermopile

In a circuit consisting of two different conductors if the two junctions are kept at different temperatures, a current will be generated, or if one of the junctions is kept open, a potential difference is developed. Through a circuit with an external load the current can produce useful work. (Fig. 2.3)

A number of such thermocouples can be connected in series to form a thermopile. The hot junction can be heated by a flat plate solar collector. (Fig. 2.4)

The potential difference with metal thermocouples is very small, up to 60 μV/degC. Using semiconductor materials (eg silicon or germanium), it can be increased to 300–400 μV/degC. The materials must be selected to suit the operational temperature range. With high currents the effect is reduced.

Using a flat plate collector to heat the hot junction and water cooling for the cold junction (which would give 50–60°C hot water as a by-product), efficiencies in the order of 1% can be achieved. For this temperature range a bismuth-antimony alloy coupled with zinc antimonide would be suitable.

2/a
Photo-emissive generators

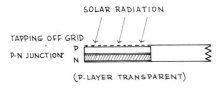

PHOTO-EMISSIVE GENERATOR

2.5
photo-emissive generator

If a high energy photon of the solar radiation strikes an atom of the cathode, the energy level of an electron may be increased sufficiently to escape from the cathode surface. If there is a suitable anode nearby, the electrons may be picked up and a current may be set up through an external circuit. As the electron emission takes place on the surface exposed to radiation, the anode must be made transparent eg a fine wire mesh. (Fig. 2.5)

Only short wave radiation (less than 620 nm) photons will have sufficient energy to achieve this effect, thus only less than 30% of the solar radiation is effective. The theoretical maximum efficiency is 15% (ie half of the available is converted) but the best so far achieved is less than 0·5%.

2/b
Photo-galvanic generators

These are actually ordinary electro-chemical cells (batteries) exposed to an additional photoelectric effect. In a galvanic battery the potential difference changes, when one of the electrodes is exposed to light. Galvanic cells consist of two electrodes (usually of different metals) submerged in an electrolyte of an acid, base or metallic salt solution. Exposure of one of the electrodes to light will induce an ion flow through the electrolyte and create a potential difference between the electrodes (or increase an already existing one).

Only high energy photons (some 45% of the solar radiation) will be effective, ie the ultra-violet and some of the visible spectrum, but not the infra-red. There are innumerable potentially suitable electrode reactions, but none seem to be very promising. Best efficiencies so far achieved are below 1%. This direction of research has been neglected in recent years, in favour of the next device.

2/c
Photo-diodes or photo-voltaic cells

SOLAR RADIATION

TAPPING OFF GRID

P-N JUNCTION P N

(P-LAYER TRANSPARENT)

2.6
photo-voltaic cell

Certain semiconductor materials can be 'doped' with minute quantities (about one part per million) of other similar elements but having one more or one less electron than the semiconductor itself. The first will be referred to as an N-type and the second as a P-type semiconductor.

eg
silicon + arsenic — N-type — one extra electron
silicon + boron — P-type — one missing electron ('hole')

If thin layers of the two are sandwiched, forming a diode, electrons will cross through the P – N junction when it is exposed to radiation. (Fig. 2.6)

Only photons above the energy threshold will be able to create such an electron/hole pair. With silicon this threshold is 1·1 eV corresponding to a wavelength of 1100 nm. Longer wavelengths will only create a heating effect.

The open circuit potential difference is around 0·6 V, and the short circuit current produced may be in the order of 0·02 A/cm². The power produced can be around 16% of the solar radiation intensity, but 18% has also been reported. The theoretical maximum efficiency is approx 24%.

With overheating the output decreases quite rapidly, thus it is advantageous to remove the unwanted heat by some form of cooling (this itself can be utilised).

Unfortunately the single-crystal cells giving the above performance, are very expensive. Polycrystalline cells cost only about one-third of the above, but their efficiency is reduced by a factor of four. A new development is the 'single cell grown ribbon' technology, which promises to reduce the cost of the most efficient device by a factor of ten.

Cadmium sulphide (CdS) is another promising material. Other semiconductors that can be used for solar cells are germanium, gallium arsenide, cadmium telluride and many others. [4] (Fig. 2.7)

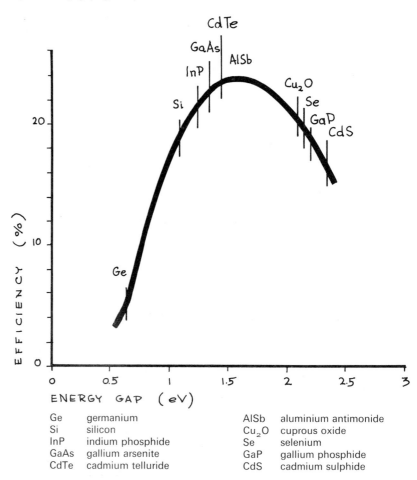

2.7
efficiency of photo-voltaic cells [4]

Ge	germanium	AlSb	aluminium antimonide
Si	silicon	Cu_2O	cuprous oxide
InP	indium phosphide	Se	selenium
GaAs	gallium arsenite	GaP	gallium phosphide
CdTe	cadmium telluride	CdS	cadmium sulphide

2.3
Thermal conversion

When radiant energy falls on a matt black surface, much of it is absorbed. This may be a complex process, which varies with the type of absorber material. It involves scattering, photon absorption, acceleration of electrons, multiple collisions, but the end effect is that the radiant energy of all grades (all wavelengths) is degraded to heat. Molecules of the surface will be excited, a temperature increase is caused. The absorption coefficient of various types of black absorbers varies from 0·8 to 0·98 (the remaining 0·2 or 0·02 is reflected).

Some of this molecular movement (ie heat) is transmitted to other parts of the body by conduction and some of it is re-emitted to the environment by convective and radiant processes. This emission of heat (heat loss) depends on the difference in temperature between the surface and the environment. Thus, as the surface is heated, the heat loss is increasing. When the rate of radiant heat input is equalled by the heat loss, an equilibrium temperature is reached.

When $Qi = Ql$

ie: $I \times a = f \times \Delta t$

the equilibrium temperature is

$$\Delta t = \frac{I \times a}{f}$$

where Qi = heat input rate (W/m²)
Ql = heat loss rate (W/m²)
I = incident intensity (W/m²)*
a = absorption coefficient
f = film, or surface conductance for emission (W/m² degC)
Δt = temp. increase over ambient
(Δ normally denotes a difference or change)

* I is the simplified term used, meaning the intensity denoted as I_{pt} in section 1.9.

The value of f depends on the material, on its surface texture, on air velocity passing the surface and on the temperature of surfaces opposite the absorber (at any distance); thus it allows for both convective and radiant heat transfer processes. Beyond the normal (0°–40°C) range it is no longer constant. We shall come back to this later (in section 4.1). For the moment it is sufficient to assume that

$$f = 11 + 0.85 \times v \qquad \text{where } v = \text{air velocity (m/s)}$$

thus with a very light air movement (just over 1 m/s)

$$f = 12 \text{ W/m}^2 \text{ degC}$$

With an intensity of 400 W/m² and absorption coefficient of 0.9, the equilibrium temperature will be

$$\Delta t = \frac{400 \times 0.9}{12} = 30 \text{ degC above air temperature,}$$

that is if there is no conduction of heat away from the back of the absorber surface.

In a given time (eg one hour) the temperature increase can be found from

$$Hi = Hl + Hg \qquad \text{where } Hg = \text{heat gain (Wh)}$$
$$Hl = \text{heat loss (Wh)} = Ql \times h \times A$$
$$Hi = \text{heat input (Wh)} = Qi \times h \times A$$
$$h = \text{number of hours}$$
$$A = \text{area (m}^2\text{)}$$

But

$$Hi = I \times a \times h \times A$$
$$Hl = f \times \Delta t \times h \times A$$
$$Hg = \Delta t \times C$$

where C = heat capacity of body (Wh/degC) (mass × specific heat, or volume × volumetric specific heat)

taking 1 hour and 1m² the h and A terms may be omitted, thus by substituting we get

$$I \times a = f \times \Delta t + \Delta t \times C$$

from which

$$\Delta t = \frac{I \times a}{f + C} \qquad \text{(f W/m}^2 \text{ degC is multiplied by 1 h and 1 m}^2\text{, thus dimensionally it represents Wh/degC in this case)}$$

With the conditions of the above example, if we take a 1m² steel radiator panel of 5 kg mass
(specific heat: 0.13 Wh/kg degC)
containing 1.5 litre water
(volumetric specific heat: 1.16 Wh/lit degC)
thus C = 5 × 0.13 + 1.5 × 1.16 = 2.39 Wh/degC

in one hour it will show a temperature rise of

$$\Delta t = \frac{400 \times 0.9}{12 + 2.39} = 25 \text{ degC}$$

Thus Hi is 360 Wh, of which the loss is Hl = 300 Wh and the heat gained is Hg = 60 Wh.

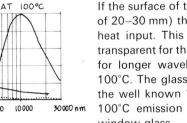

2.8
emission curves at solar and terrestrial temperatures compared with the transmission curve of ordinary window glass

If the surface of the absorber plate is covered by a sheet of glass (with an air space of 20–30 mm) the heat loss is significantly reduced without much reduction in the heat input. This is due to the selective transmittance of the glass. It is highly transparent for the short wave, high temperature solar radiation, but virtually opaque for longer wavelength infra-red radiation emitted by the absorber plate below 100°C. The glass cover also reduces the convective losses from the plate. This is the well known 'greenhouse effect'. Fig. 2.8 shows the 6000°C solar emission, a 100°C emission spectrum and the spectral transmission curve for an ordinary window glass.

The glass will cause some reduction of the radiation intensity on the absorber plate, ie there is an optical loss in transmission, but this is much less than the resultant saving on heat loss. The proportion transmitted is expressed by the transmission coefficient. This has a constant value for diffuse radiation, but for direct radiation it is a function of the angle of incidence. Some typical values for an ordinary window glass are given in Table 2.1 (average of several published sets of values).

Table 2.1 θ = Transmission coefficients

	for diffuse	for direct, if angle of incidence is						
		0°	20°	40°	50°	60°	70°	80°
single	0·70	0·80	0·80	0·79	0·77	0·72	0·60	0·38
double	0·62	0·75	0·75	0·72	0·68	0·60	0·48	0·28

When calculating the intensity of radiation incident on a tilted plane (cf section 1.9), the appropriate transmission coefficients should be applied to the diffuse and direct components, before they are added.

Having a glass cover, the heat loss from the plate will be almost exclusively convective-conductive, thus the 'air-to-air transmittance' concept (the U-value) used in building heat loss calculations will be applicable.

This can be taken as U = 5·00 W/m² degC for single glass
and U = 2·70 W/m² degC for double glass

For low temperature applications the heat loss rate can be taken as

Ql = U × Δt

and the total heat loss as

Hl = U × Δt × h × A

Substituting these values in the above example, we get

$$\Delta t = \frac{400 \times 0 \cdot 9 \times 0 \cdot 8}{5 + 2 \cdot 39} = 39 \text{ degC for single glass}$$

$$\Delta t = \frac{400 \times 0 \cdot 9 \times 0 \cdot 75}{2 \cdot 7 + 2 \cdot 39} = 53 \text{ degC for double glass}$$

Some authors suggest that when the value of Δt exceeds about 20 degC, it will be more accurate to use the non-linear relationship

$$Ql = c \times \Delta t^{1 \cdot 25}$$

where c = 2·38 W/m² degC for single glass
and c = 1·70 W/m² degC for double glass

2.4 Flat plate collectors

If some thermal fluid (eg water or air) is circulated as a carrying medium in thermal contact with the absorber plate, it will be heated and thus some of the heat absorbed by the plate will be removed. The temperature of the plate is thereby reduced to below the above calculated equilibrium temperature, and this will reduce the heat loss.

The equation for temperature rise is still valid, but the thermal capacity term (C) should include the amount of water flowing through in the time considered, in lieu of the amount contained in the panel. If we continue the above example, but assume a flow rate of 8 litre/h, thus

C = 5 × 0·13 + 8 × 1·16 = 9·93 Wh/degC, we get

$$\Delta t = \frac{400 \times 0 \cdot 9 \times 0 \cdot 8}{5 + 9 \cdot 93} = 19 \text{ degC with single glass}$$

$$\Delta t = \frac{400 \times 0 \cdot 9 \times 0 \cdot 75}{2 \cdot 7 + 9 \cdot 93} = 21 \text{ degC with double glass}$$

These would be the increases in temperature over the ambient, if it is assumed that the starting temperature and the cold water inlet temperature are the same as the ambient. Thus if we take for example 15°C as the starting temperature, from this particular collector we would get 8 litres of water in one hour at 15 + 19 = 34°C if it was single glazed and at 15 + 21 = 36°C if it was double glazed.

The above is valid for the starting condition. In normal operation we must distinguish two temperature differences:

1   for effective gain: $\Delta t = t_f - t_r$
where t_f = flow temperature from plate
t_r = return temperature to plate
ie the increase in water temperature through the plate from 'return' to 'flow'

2   for heat loss: $\Delta t = \dfrac{t_r + t_f}{2} - t_o$

ie the mean plate temperature minus the outdoor air temperature.

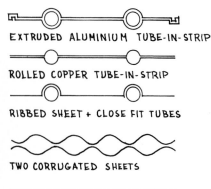

EXTRUDED ALUMINIUM TUBE-IN-STRIP

ROLLED COPPER TUBE-IN-STRIP

RIBBED SHEET + CLOSE FIT TUBES

TWO CORRUGATED SHEETS

FLAT + RIBBED SHEET

2.9
absorber plate sections

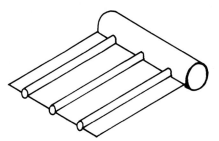

TUBES WELDED INTO LARGER
HEADER PIPE

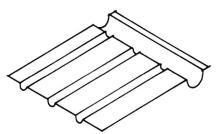

LOWER SHEET PRESSED, INCL.
HEADER + FLAT TOP SHEET

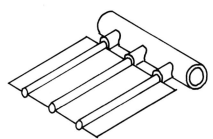

TUBES CONNECTED TO
RUBBER HEADER

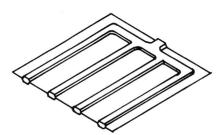

ALUMINIUM ROLL-BOND PANEL

2.10
arrangement of headers

CLOSELY SPACED BLACK
PLASTIC TUBES

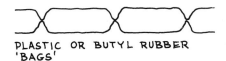

PLASTIC OR BUTYL RUBBER
'BAGS'

RIGID PLASTIC TRAY WITH
FLAT METAL TOP

2.11
plastic absorbers

We can substitute into the equation

$$H_g = H_i - H_l$$

the following values:

heat gain: $H_g = \Delta t \times C = (t_f - t_r) \times C$

> as there is no change in plate temperature the thermal capacity of the water flowing through is only to be considered.
> If calculated for 1 hour
> $C = F \times s$
> where F = flow rate (litre/h)
> s = volumetric specific heat of water = 1·16 Wh/lit degC

heat input: $H_i = I \times A \times \theta \times a$

> where I = incident intensity (W/m² = Wh/m²h)
> A = area of plate (m²)
> θ = transmission coefficient of glass
> a = absorption coeff. of plate

heat loss: $H_l = A \times U \times \Delta t = A \times U \times \left(\dfrac{t_r + t_f}{2} - t_o \right)$

thus we get

$$C(t_f - t_r) = I \times A \times \theta \times a - A \times U \times \left(\dfrac{t_r + t_f}{2} - t_o \right)$$

If all other values are known, we can determine t_f rearranging the equation

$$t_f = \frac{IA\theta a + t_r \left(C - \dfrac{AU}{2} \right) + AUt_o}{C + \dfrac{AU}{2}}$$

This value can be substituted into the heat gain expression above to get the amount of useful collection.

The absorber plate itself can be any metal sheet, incorporating water channels. In the simplest case it may be an ordinary central heating radiator panel. Many steel, copper and aluminium products are on the market which may be suitable. (Fig. 2.9)

The water channels must be connected at top and bottom by some form of header or manifold.

The header should have a cross sectional area larger than the aggregate area of the channels served, to ensure a balanced and uniform flow in all channels. (Fig. 2.10)

Non-metal absorbers are also possible, but their profile will necessarily be different. Much closer contact between the surface and the liquid will be necessary, as the thermal conductivity of plastics and rubber is much less than of metals. Some possibilities are shown in Fig. 2.11.

Surface finish of the absorber plate may be a matt black paint, such as a 'chalkboard black', with an appropriate rust inhibiting primer. The latter should preferably be a wash-primer, as a thick undercoat of paint would reduce the transmission. The primer should be of the self-etching type. Without this the repeated thermal expansion and contraction of the plate may cause the paint to peel off after a year or so. Several types of baked-on finishes are also available.

The so-called 'selective surfaces' have a high absorption and emission coefficient for the 200–2000 nm solar radiation, but a much lower a and e value for the longer infra-red (up to 20 000 nm) emitted by bodies at a temperature below 100°C. A measure of their performance is the a/e ratio, ie the ratio of absorption coefficient for solar radiation, to the emission coefficient at operating temperatures. A number of such surfaces have been tested or developed. They fall into three basic categories: [5]

1 *multi-layer interference coating*, consisting of several layers of different materials in strictly controlled thickness. Two examples and their performance are given in Fig. 2.12.

2 *bulk absorbers* mostly of copper oxides with an a/e ratio of 6 are commercially available. Higher performance bulk absorbers are still in an experimental stage.

3 *meshes* (or multiple cavity absorbers) can be formed by laser beam interference on a photosensitive surface, which can subsequently be covered with a vacuum

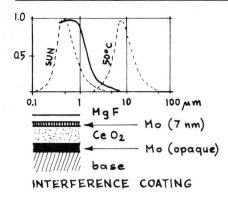

MgF
Mo (7 nm)
CeO₂
Mo (opaque)
base

INTERFERENCE COATING

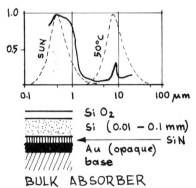

SiO₂
Si (0.01 – 0.1 mm)
SiN
Au (opaque)
base

BULK ABSORBER

2.12
selective surfaces showing the
absorption/emission curves compared
with solar and terrestrial radiation spectra

2.13
efficiency of collectors (after E Speyer in
New Sources of Energy, UN Conf, Rome,
1961)

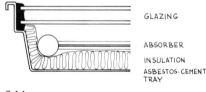

2.14
a collector panel

deposited aluminium film. So far this has been done only on a laboratory scale, but the results are promising. Reflection is very low for wavelengths up to 1·3 times the mesh constant, then it rapidly rises to 0·95 at a wavelength of 2 × mesh constant.

Most of the work on selective surfaces is still experimental. Baum [6] reports the development of an absorber ($Ni + SiO + MbF_2$) achieving an absorption coefficient of 0·90 and emission of 0·05, ie an a/e ratio of 18. Seraphin and Wells produced a surface [7] by chemical vapour deposition producing an a/e ratio up to 28.

The most generally used selective surfaces are either oxidised copper or particulate copper oxide deposition on some other material. Generally, the use of these rather expensive surfaces is justified only in case of high temperature collection (over about 60°C). Fig. 2.13 shows the efficiency of collectors with ordinary black and with selective surfaces, with 1, 2 and 3 glass covers. [8]

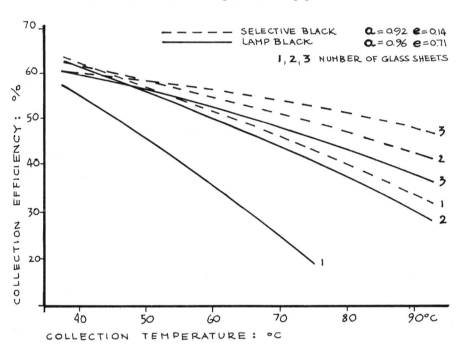

The absorber plate may be enclosed in a tray of some kind (asbestos-cement, fibreglass, metal or wood) incorporating a substantial thermal insulation at the back and edges (50–100 mm) as well as holding the glass front. The complete unit is usually referred to as the 'collector panel'. (Fig. 2.14)

These ready-made units are normally used for small area water heaters. For large areas the absorber plate may be independent of the glass and its framing, and both may be incorporated with the fabric of the building, replacing a wall or a roof.

If air is used as the collection fluid, the construction of the collector will necessarily be different. Firstly a far larger volume flow rate will be necessary. The specific heat of 1 m³ of air is around 0·36 Wh/degC whilst that of 1 m³ of water is 1160 Wh/degC. Secondly the coefficient of heat transfer from a surface to air is much less that to a liquid. Therefore the collector will have to be much bulkier and must incorporate a large heat transfer area. Several systems have been developed, but the possibilities are not nearly as well explored as with waterborne collection systems. (Fig. 2.15)

2.5 Performance

Efficiency of the solar collector depends on a number of factors and will be greater if

1 glass transmission coefficient is maximised
2 transmittance for outgoing heat flow is minimised
3 absorption coefficient of plate is maximised
4 emission coefficient for long wave is minimised
5 plate temperature is kept to the minimum useable level, as this will minimise the heat loss.

The term *plate efficiency* relates the actual performance to the possible maximum: 100% would be achieved if the whole surface were at a uniform temperature, this being the same as the water temperature. It will be reduced by two effects:

1 half-way between the water channels the temperature of the plate will be higher than immediately above the water, thus a temperature gradient is developed. (Heat input is uniform over the whole surface, heat removal takes place in the

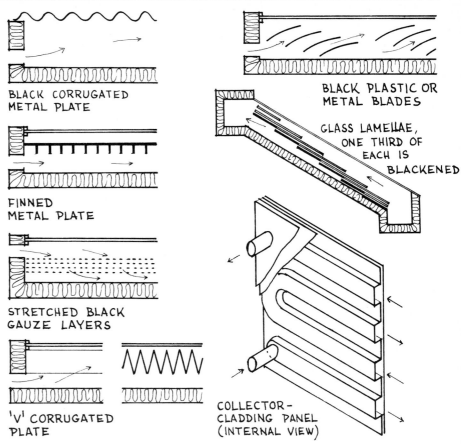

BLACK CORRUGATED
METAL PLATE

FINNED
METAL PLATE

STRETCHED BLACK
GAUZE LAYERS

'V' CORRUGATED
PLATE

BLACK PLASTIC OR
METAL BLADES

GLASS LAMELLAE,
ONE THIRD OF
EACH IS
BLACKENED

COLLECTOR-
CLADDING PANEL
(INTERNAL VIEW)

2.15
air heaters

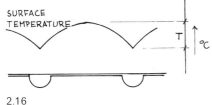

2.16
plate temperature gradient

water channels only.) The peak temperature increases the heat loss from the plate, without increasing the heat transfer to water. (Fig. 2.16) The gradient and its amplitude (T) can be reduced by a material of better conductivity, by a greater plate thickness giving a better conductance and by closer spacing of the tubes

2 the water temperature will always be less than the plate temperature, but this difference is reduced with an improvement of the heat transfer coefficient from plate to water. Turbulent rather than laminar flow will give an improvement.

Experimental results have produced the following plate efficiencies: [9]

Table 2.2 Plate efficiencies

plate material and thickness		tube spacing (mm)					
	(mm)	75	100	125	138	150	175
copper	0·25	94·5	92	89	87	85·5	80·5
	0·35	95	92·5	90	88	87	82·5
	0·45	95·5	93	91	89	88	85
	0·55	96	93·5	91·5	90	89	86·5
	0·70	96	93·5	92	91	90	87·5
aluminium	0·50	94·5	92	88·5	87	86	82·5
	0·75	95·5	93	90·5	89	88	85
	1·00	95·5	93·5	91·5	90	89	87
steel	0·50	89	82·5	75	71·5	68	62·5
	0·75	92	87	80·5	77	74	68·5
	1·00	93	88·5	83·5	81	77·5	72·5
	1·50	95	89	84	81	79·5	74·5

The term *collection efficiency* means a ratio of the actual heat collected to the amount of radiation falling on the glass, expressed as a percentage

$$\eta = \frac{Hg}{I \times A} = \frac{\text{heat gain}}{\text{incident radiation}}$$

Depending on the period considered, we can speak of 'daily', 'annual' or 'instantaneous' efficiency of collection. The magnitude of this depends on the radiation regime and external temperatures as well as on the actual collector design, water flow rate and collection temperature. [10]

Fig. 2.17 shows the daily efficiencies obtained in London, and the dependence of this on collection temperature.

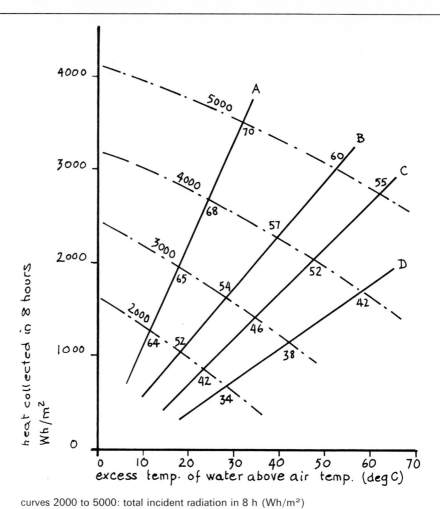

curves 2000 to 5000: total incident radiation in 8 h (Wh/m²)
figures at intersections: percentage of efficiency
A—water capacity 96 kg/m², thermal and forced circulation
B—water capacity 48 kg/m², total content at max. temperature
C—water capacity 48 kg/m², part of content at max. temperature
D—water capacity 24 kg/m², absorber only filled, static

2.17
daily collection efficiencies [10]

As a rough guidance, for a double glazed non-selective black absorber plate the annual collection efficiency can be taken as 0·4 times the plate efficiency given above.

2.6 Improvements

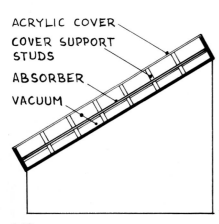

2.18
a vacuum absorber

One way of improving the performance of flat plate collectors is to reduce the heat loss by creating a partial vacuum between the absorber plate and its transparent cover. To prevent the collapse of this transparent cover under atmospheric pressure when the inside is evacuated, short spacer-studs (eg of perspex) can be inserted. Similar studs may keep the absorber plate and the backing sheet apart. The vacuum will be a better insulator than any practicable thickness of porous insulating material. Fig. 2.18 shows the diagrammatic cross-section of such an absorber and Fig. 2.19 indicates the improvements that can be expected with two different levels of vacuum. The improvement is more pronounced with selective surfaces. In this case radiant losses are reduced by the low emittance of the absorbing surface and the conductive losses are diminished by the vacuum between absorber and its covers. [11]

A variety of this idea is an all-glass unit, to be manufactured in long plank form in widths of 200–300 mm. The top chamber is a partial vacuum to reduce losses. The bottom chamber is similar, but the inside surfaces are silvered to reduce emission and provide reflective insulation. The middle chambers carry the water flow under normal operating pressures. Fig. 2.20 shows such a collector, which could be produced for about £13/m².

When it is necessary to produce higher temperatures, the heat loss from the plate becomes increasingly significant. Double glazing and a selective absorber surface will have to be used. Both are rather costly. One way of economising is to use a multi-stage collector. [12] An inexpensive ordinary black, single glazed unit to act as a pre-heater, a similar but double glazed unit as a second stage, followed by a third stage with both double glazing and selective surfacing. (Fig. 2.21)

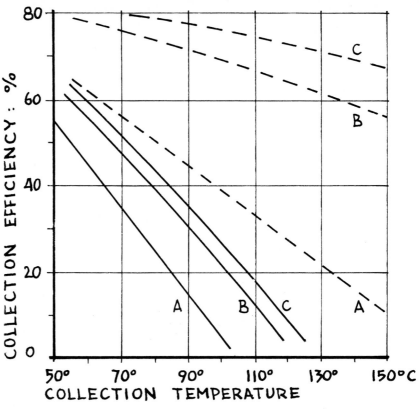

COLLECTION EFFICIENCY : %

COLLECTION TEMPERATURE

- - - - SELECTIVE BLACK

———— NON-SELECTIVE BLACK

A   ATMOSPHERIC PRESSURE $(1.013 \times 10^5 \, N/m^2)$

B   WITH $3385 \, N/m^2$

C   WITH $133 \, N/m^2$

2.19
performance of a vacuum absorber

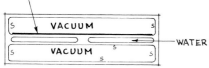

SELECTIVE BLACK

S | VACUUM | S
VACUUM ——WATER

S = SILVERED SURFACES

2.20
an all-glass vacuum absorber

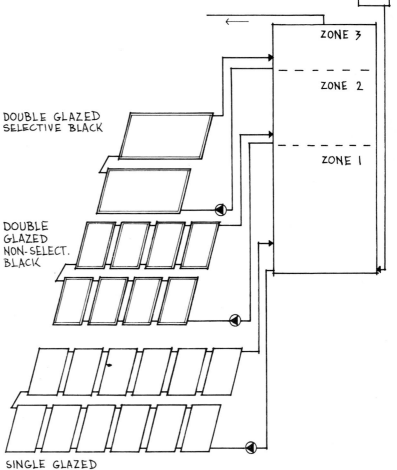

DOUBLE GLAZED
SELECTIVE BLACK

DOUBLE
GLAZED
NON-SELECT.
BLACK

SINGLE GLAZED

ZONE 3

ZONE 2

ZONE 1

2.21
a multi-stage collector

2.7 Alternatives

Any dark surface is an absorber. The removal of heat through closed flow absorber panels with a thermal fluid (water or air) is only one way of utilisation. There are four further possibilities: **a** open flow collectors, **b** solar ponds, **c** container heaters, **d** buildings as collectors.

a   A dark surface can have a film of water running down on it, under a transparent cover. This system can be almost as efficient as a closed flow absorber at much less cost.

b   The bottom of a shallow pool may be made dark. Some of the solar radiation will be absorbed by the water, whilst passing through, then some by the dark bottom. Any energy reflected from the bottom will be further reduced by absorption in the water on the way back. The method is quite successful for low temperature collection. It is also used in some solar houses, to be shown in part 4. In such ponds the bottom layer of water is heated the most. It will rise to the surface and thus convection currents will develop. The warmest layer settling at the top will increase the heat loss. If however a salt solution is used, the warmer water will support more salt, thus in spite of its temperature it will settle at the bottom. In a 1 m deep and 25 × 25 m pond the bottom temperature has reached 93°C.

c   Container heaters may consist of a shallow metal tank or tray with a transparent cover or they may take the form of a *solar cushion*. (Fig. 2.22) They constitute a combination of the absorber and the storage. Their common feature is that there is

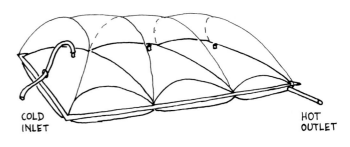

TRANSPARENT VINYL COVER

COLD INLET

HOT OUTLET

HOT WATER CUSHION, BLACK VINYL
(1.8 × 0.9 × 0.12 m = 200 LITRE CAPACITY)

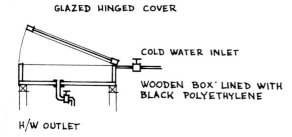

GLAZED HINGED COVER

COLD WATER INLET

WOODEN BOX LINED WITH BLACK POLYETHYLENE

H/W OUTLET

2.22
low-cost water heaters, a 'hot-box' and a 'cushion' type

no flow, a given amount of water is heated, then used, when the content of the heater will be replaced by fresh, cold water.

d   Buildings designed to act as solar collectors will be discussed in part 4.

2.8 Heat pumps

Very significant amounts of energy are available as low grade heat, at temperatures too low to be useful for any practical purpose. Heat pumps may be utilised to up-grade such heat to useful temperatures.

The heat pump is a refrigeration machine used in reverse. A *refrigerant,* such as Freon, methyl chloride or ammonia is circulated in a closed cycle by a compressor. A pressure release (or choke) valve keeps it under pressure at the hot end (in the condenser) and under a reduced pressure at the cold end (in the evaporator). (Fig. 2.23) When compressed, the temperature of the refrigerant is increased and it will liquefy whilst giving off heat into the 'sink'. Passing through the pressure release valve, it rapidly evaporates and drops in temperature, taking up heat from its environment, from the 'source'.

The coefficient of performance of heat pumps is

$$\frac{Q}{W} = \frac{\text{heat delivered to sink}}{\text{work input}}$$

In the ideal (Carnot) cycle this coefficient of performance is inversely proportionate to the temperature difference

$$\frac{Q}{W} = \frac{T'}{T' - T''} \qquad \begin{array}{l} \text{where } T' = \text{sink temperature} \\ \phantom{\text{where }} T'' = \text{source temperature} \end{array} \Big\} \text{ in } °K$$

(see Fig. 2.24)

The practical (Rankine) cycle will give a value of some 0·8 times the above, but this is further reduced by the component efficiency factors, such as

2.23
heat pump, showing its principles and an
actual system

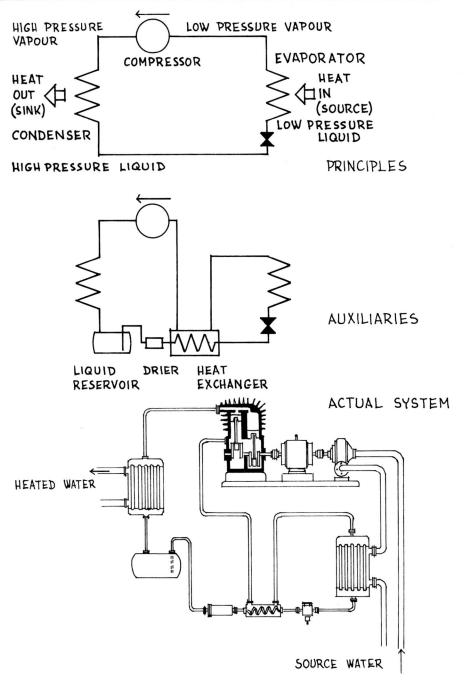

electric motor driving the compressor: $\eta_{m} = 0.95$
compressor: $\eta_{c} = 0.80$
heat exchangers: $\eta_{h} = 0.90$

The overall efficiency will be a product of these:

$\eta = 0.8 \times 0.95 \times 0.8 \times 0.9 = 0.55$

Thus eg to transfer heat from a source of 10°C (283°K) to a sink at 50°C (323°K)
we get a coefficient of performance

$\frac{Q}{W} = 0.55 \frac{323}{323 - 283} = 4.8$

Thus with a 1 kW compressor motor we can expect to deliver heat at a rate of
4·8 kW. As the table overleaf shows, existing heat pump installations have achieved
coefficients of performance between 3 and 6. [13]

Heat pumps can be used directly for solar energy collection. If the evaporator is the
absorber plate, it becomes the 'source'. The condenser may be a heat exchanger in
a water tank, and act as a water heater. (Fig. 2.25)

In the indirect system a tank of water may be the 'source' which is cooled by the
evaporator. This cooled water is then circulated to the collector plate where it will
be heated and returned to the tank. Thus with the reduced collection temperature
the collection efficiency is greatly improved. (Fig. 2.26)

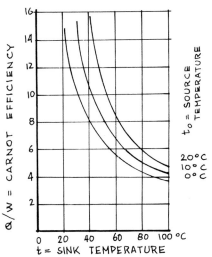

2.24
heat pump efficiencies

Table 2.3 Heat pump installations in the UK

location	bldg. type	max. output kW	source	c.o.p.
Norwich	office block	135	river	3·0
Stourport (Worcs)	workshops	220	cooling water	4·8
Gt. Yarmouth	office block	490	cooling water	3·8
London	Festival Hall	2700	Thames	3·0
Oxford	college	147	sewage	4·0
Shinfield (Berks)	laboratory	28	soil	3·4
Birmingham	commercial	24	canal	3·5
Meaford (Staffs)	offices	135	cooling water	5·2

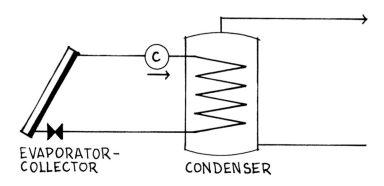

2.25
direct solar heat pump

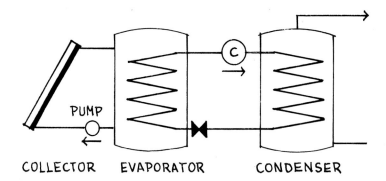

2.26
indirect solar heat pump

Several of the solar houses described in part 5 rely, at least partially, on some form of a heat pump system.

2.9
Concentrating devices

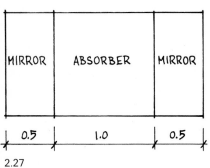

2.27
two-mirror concentrator

It has been shown that the temperature of absorbent surfaces exposed to solar radiation will increase until an equilibrium condition is reached. In section 2.3 this equilibrium temperature has been calculated for an incident radiation of 400 W/m² as

exposed surface	25 degC above ambient
one glass cover	39
two glass covers	53

If higher temperatures are required, some form of optical concentration of the radiant energy will be necessary. Haywood suggests [10] that with an incident radiation of 630 W/m² the following equilibrium temperatures can be produced:

flat plates:	one glass cover	59 degC above ambient
	two glass covers	71
concentrators:	concentration rate	
	5	178 degC above ambient
	10	306
	20	520

Concentration rate (Cr) is the ratio of the area exposed to the sun, normal to the solar beam (ie the 'catchment area') to the area of the solar image produced by the device on the absorber.

The advantage of concentrating devices is not only the higher temperature produced, but also the fact that whilst heat is gained from a large area, heat loss takes place only from a small surface (the actual absorber). Their disadvantage is that they can utilise only the direct, directional radiation, they will not respond to diffused radiant energy.

Five basic types of concentrating devices can be distinguished:

1 plane mirrors
2 parabolic troughs (cylindrical reflectors)
3 paraboloid reflectors
4 cylindrical Fresnel lenses
5 circular Fresnel lenses

Lenses can achieve a great optical accuracy but beyond a few hundred mm diameter they are very expensive. They are only used in conjunction with reflectors as secondary concentrators. Reflectors and mirrors however deserve a brief review.

2.10
Plane mirrors

The simplest concentrating device is an absorber plate with two plane mirrors of the same size, set at an angle of 60° (this may be a hinged, fold-out construction). If the absorber and mirrors are 1 m² each, the projected (catchment-) area is 2 m², thus the concentration rate is $Cr = \frac{2\ m^2}{1\ m^2} = 2$ (Fig. 2.27).

With four such mirrors the exposed area is 3 m², thus $Cr = \frac{3\ m^2}{1\ m^2} = 3$ (Fig. 2.28).

The effective concentration rate ($C\eta$) is somewhat less, as any mirror is a less than perfect reflector and the absorption of the plate is reduced with the low angle of incidence. However, with the four-mirror arrangement shown above $C\eta = 2$ can be achieved even with crude devices.

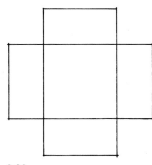

2.28
four-mirror concentrator

Many such plane mirrors of identical sizes can be set up each directing its reflection on to one common absorber plate.

2.11
Parabolic troughs

Parabolic trough reflectors have a line-focus. As however the rays arriving from the solar disc at a particular point are convergent, the reflected rays will be divergent, thus even with an optically perfect reflector the sun's image at the focal line will not be a point, but will have a certain width.

The angle subtended by the sun's radius (γ) is 16' (or 0·00465 radians), the image will have a radius of 0·00465 d, where d is the distance from a point of the parabola to the focus (Fig. 2.29).

In triangle PFK or PFL the angle at P is 0·00465 rad

the distance $\overline{PF}$ is = d thus the side $\overline{KF}$ will be 0·00465 d or $\tan 16' = \frac{\overline{KF}}{d}$ from which

$\overline{KF} = \tan 16'\ d = 0·00465\ d$ (for small angles the tangent is the same as the angle in radians, ie as the length of arc at unit distance) thus the image diameter $\overline{KL}$ will be 0·0093 d.

In practice this can be taken as 0·01 d. [14]

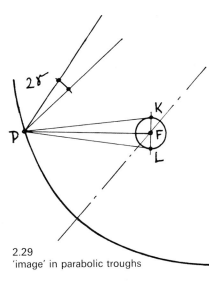

2.29
'image' in parabolic troughs

The parabolic reflector can be characterised by the rim angle ϕ. (Fig. 2.30) The width of the catchment area is 2 b and b = sin ϕ d. Thus the catchment area is 2 sin ϕd × length. With ϕ = 90° the catchment width is 2 d, thus the concentration rate is $Cr = \frac{2\ d}{0·01\ d} = 200$.

Fig. 2.31 shows three methods of constructing a parabola:

I   if vertex V and focus F are given on axis a
— mark point D so that $\overline{DV} = \overline{VF}$
— draw lines perpendicular to the axis at any points 1, 2, 3, 4, 5, etc
— bisect each with an arc drawn from F, with a radius equal to the distance of the particular point from D (thus bisect line 3 with radius $\overline{D3}$ drawn from F)

II if (as above) vertex V and focus F are given on axis a
— draw line from V perpendicular to axis
— mark any point on this line (1, 2, 3, etc) and connect it to F
— draw a line from each point at right angles to the line to F (thus from point 5 draw line at right angles to $\overline{5F}$)
— each of these lines will be a tangent of the parabola, which must be drawn inside these tangents, touching each section at its centre

III if vertex V and axis $\overline{VA}$ are given as well as width $\overline{BA}$
— draw $\overline{CB}$ parallel to axis and $\overline{CV}$ perpendicular to axis
— divide both into an equal number of equal sections (1, 2, 3, 4)
— connect V to each point of $\overline{CB}$ (radial lines)
— intersect each radial line with parallels drawn from corresponding points of $\overline{CV}$.

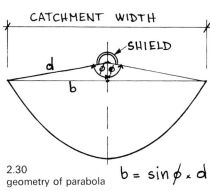

2.30
geometry of parabola   b = sin ϕ × d

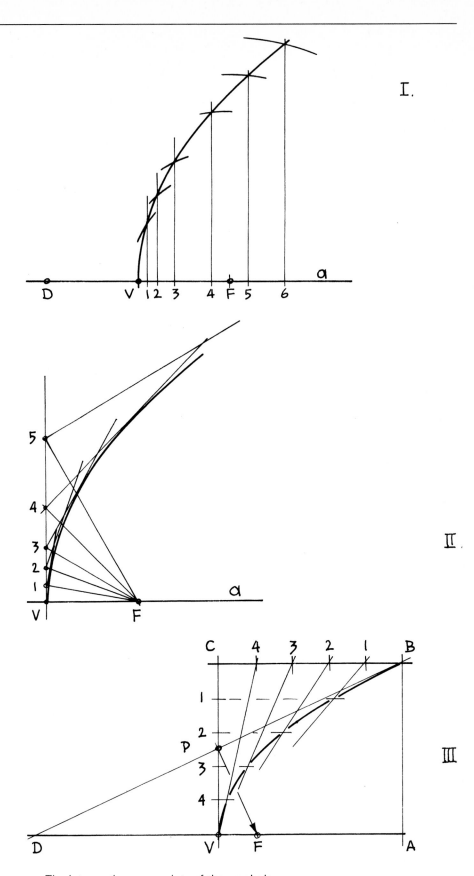

I.

II.

III.

2.31
construction of parabola

The intersections are points of the parabola
— mark point D on extension of axis, so that $\overline{DV} = \overline{VA}$
— the line connecting D with B is a tangent at B
— where $\overline{DB}$ intersects $\overline{CV}$, mark point P
— draw line from P at right angles to $\overline{BD}$, this will determine the position of focus F on the axis

This type of collector may be mounted on a horizontal east-west axis. A tracking device may adjust its tilt to follow the solar altitude angle. An absorber tube will be located along the focal line. Re-radiation from this tube should be prevented by a cylindrical shield, having a slot angle the same as the rim angle of the reflector (Fig. 2.30).

An equilibrium temperature is reached when the heat loss equals the heat input

$Q_i = I \, a \, Cr$

$Q_l = e \, \sigma \, T^4$ where e = emittence of absorber

T = temperature of absorber (°K)

σ = Stefan-Bolzmann constant ($5 \cdot 77 \times 10^{-8}$ W/m² degK⁴)

Fig. 2.32 gives the equilibrium temperatures of absorbers as a function of concentration rate. [14]

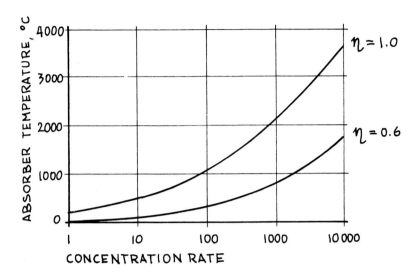

2.32
temperature vs concentration rate

2.12
Paraboloid
reflectors

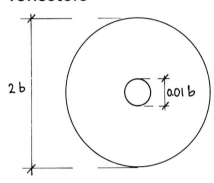

2.33
paraboloid reflector (frontal view)

The paraboloid is produced from the rotation of a parabola around its axis. Its concentration rate will be the square of the concentration rate of the parabola. Fig. 2.33 gives the frontal view of such a paraboloid. It can be seen that the catchment area is

$$\frac{(2b)^2 \pi}{4} \quad \text{and the image area is} \quad \frac{(0 \cdot 01b)^2 \pi}{4}$$

thus the ratio of the two is

$$\frac{2^2}{0 \cdot 01^2} = 200^2 = 40\,000$$

which is the theoretical concentration rate of such a paraboloid. In practice the real image has at least twice the theoretical diameter, thus the concentration rate will be about 10 000.

To be effective, these collectors must track the sun both in azimuth and in altitude, ie they must have a *heliostat* mechanism of some kind.

The removal of heat from the focal point may represent difficulties in practical engineering. This problem can be avoided by cutting a hole in the centre of the paraboloid and using a second mirror to reflect the radiation to the absorber, which can thus be placed behind the mirror. (Fig. 2.34)

We may distinguish short focus (deep) paraboloids and long focus (shallow) ones (Fig. 2.35). The latter one requires a more accurate tracking mechanism, but it can have a flat absorber disc and it is easier to achieve optical accuracy with this type. With short focus reflectors the absorber is usually hemispherical, it has some tolerance in directionality, thus the tracking need not be so accurate, but it is more difficult to achieve optical accuracy with such deep paraboloids. The best results have been achieved with reflectors having a rim angle of 90°.

If a less than perfect focusing is acceptable, a shallow spherical reflector can also be used.

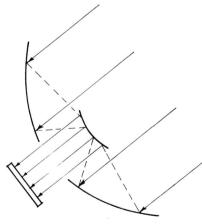

2.34
double paraboloid reflector

The central problem in the design of paraboloid reflectors is the tracking mechanism. The simplest solution is the *azimuthal mounting*, which allows rotation around a vertical axis and tilting around a horizontal axis. Both must be operated continuously to achieve tracking. Manual adjustment is easy, but motorised tracking would require a rather complicated mechanism. (Fig. 2.36)

The *equatorial mounting* gives the greatest precision. (Fig. 2.37) Axis 1 is fixed in a position parallel to the earth's axis, ie in a north-south direction, with a tilt equal to the geographical latitude. Turning around this axis gives the diurnal rotation

2.35
deep and shallow paraboloids

References

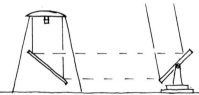

2.36
paraboloid with azimuthal mounting

2.37
paraboloid with equatorial mounting

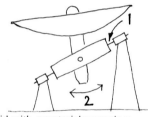

2.38
indirect tracking, equatorial

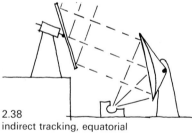

2.39
indirect tracking, azimuthal

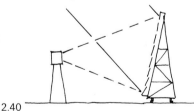

2.40
an array of plane mirrors

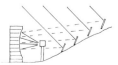

2.41
the Odeillo furnace

(15° per hour). The tilt, 2, is adjusted daily to match the solar declination (+23° to −23°).

Other systems may use a fixed paraboloid mirror, with a fully adjustable primary plane mirror. This plane mirror itself may be on an equatorial mounting (Fig. 2.38) or on an azimuthal mounting (Fig. 2.39). In the latter case there is also a second plane mirror tilted at 45° to reflect the radiation vertically upwards, so that the paraboloid reflector surface is in a much more protected (downward facing) position.

A large panel consisting of an *array of plane mirrors,* each mirror being individually adjustable, can be mounted on a semicircular rail track. All the small plane mirrors would reflect the sun's image on one absorber mounted on a tower. (Fig. 2.40)

The largest installation so far consists of over 60 arrays, each consisting of 180 plane mirrors, mounted on a hillside. Each is individually adjustable and governed by a central computer. All are reflecting on to a huge parabolic mirror (which is one side of a multi-storey office block), which in turn focuses on to an absorber, the 'furnace'. (Fig. 2.41)

1 Kormondy, E J
Concepts of ecology
Prentice-Hall, 1969
2 Odhum, H T
Environment, power and society
Wiley-Interscience, 1971
3 Brinkworth, B J
Solar energy for man
Compton Press, 1972
4 Ravich, L E
Thin film photovoltaic devices . . .
paper S 56
'New sources of energy' Conf. Rome, 1961
UN 1964
5 Australian Academy of Science
Solar energy research in Australia
report No. 17, Sept. 1973
6 Baum, V A et al.
Effet de la sélectivité dans les installations solaires énergétiques
paper E 41
'The sun in the service of mankind' Conf. Paris, 1973
7 Seraphin, B O and Wells, V A
Solar energy thermal converters fabricated by chemical vapour deposition
paper E 58
ibid
8 Speyer, E
Solar buildings in temperate and tropical climates
paper S 8
'New sources of energy' Conf. Rome, 1961
UN 1964
9 Commonwealth Scientific & Industrial Research Organisation (Australia)
Solar water heaters
Div Mech Eng. Circular No. 2
CSIRO 1964
10 Haywood, H
Solar energy for water and space heating
in J of Inst of Fuel, July 1954
11 Blum, H A et al.
Design and feasibility of flat plate solar collectors to operate at 100–150°C
paper E 18
'Sun in the service of mankind' Conf. Paris, 1973
12 Morse, R N et al.
High temperature solar water heating
paper E 60
ibid
13 Griffith, M V
Heat pump progress in Great Britain
in 'Direct Current', March 1960
14 Robinson, N
Solar radiation
Elsevier, 1966

Part 3 — Uses

3.1 Domestic water heating

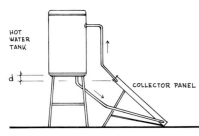

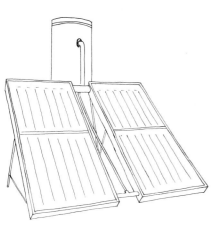

3.1
thermosyphon water heater

The simplest water heating system, based on the thermosyphon principle, consists of a collector panel and a tank. The tank must be positioned above the top of the collector panel. The water heated in the collector, becoming lighter, will rise and colder (heavier) water will be drawn in its place from the bottom of the tank. The greater the height difference (d), the larger the flow will be (as induced by the same temperature difference). A larger flow rate will increase the collection efficiency, although reduce the collection temperature. (Fig. 3.1)

In a series of experiments [1] the following efficiencies have been achieved:

if d = 0   efficiency = 0·46
    = 0·6 m   = 0·55

This is the type of system most developed. Numerous products are commercially available in many countries. For a domestic hot water installation the system usually consists of a pair of panels, 1·5–3·0 m² each, and a tank of 150–200 litre capacity. The tank itself must be well insulated and weatherproofed.

An Australian source [2] suggests that to provide 200 litres of hot water at 58°C in Melbourne (lat 32°S) about 5 m² collector area would be required in the summer and double this area (some 10 m²) in winter. With the larger area for much of the year the system would produce surplus heat, thus the law of diminishing returns applies. It may be more economical to use 6–7 m² collector and rely on an auxiliary heat source in winter for 'topping up'.

No large quantities of energy will be required, thus the most convenient auxiliary heat source is an electric immersion heater, operated on cheap off-peak electricity. This can be positioned in the collection tank itself (Fig. 3.2/a). With inclement weather, when hot water is produced by the immersion heater only, the top third of the tank only being at the desired temperature, it may not give enough hot water. If however the heater is positioned lower in the tank, it may induce a reverse flow and lead to a heat loss from the collector plate. Fig. 3.2/b shows an arrangement which would avoid this, without a complicated control gear. In this case the solar collector panel would in fact act as a pre-heater.

A somewhat more expensive, but certainly more efficient arrangement uses two

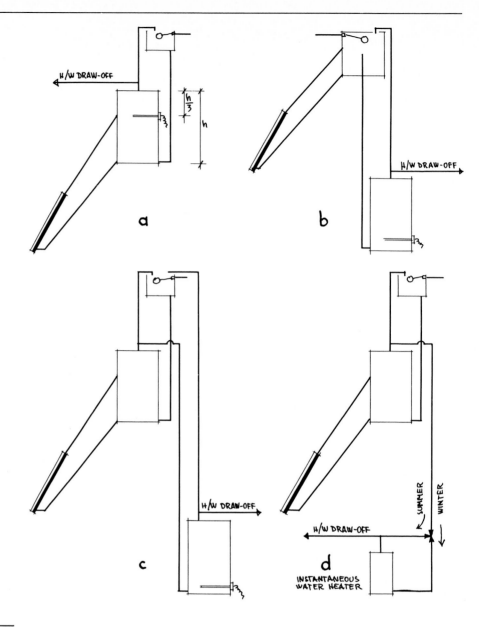

a

b

c

d

INSTANTANEOUS
WATER HEATER

3.2
solar + auxiliary water heaters

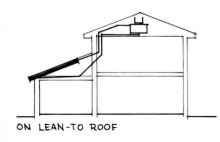

AS LOWER PART OF ROOF

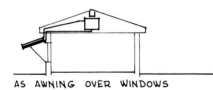

AS AWNING OVER WINDOWS

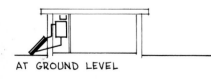

AT GROUND LEVEL

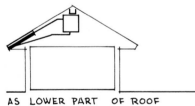

ON LEAN-TO ROOF

3.3
positioning of collectors

tanks (Fig. 3.2/c). The solar pre-heater can also be coupled with an instantaneous topping-up heater, either gas or electric (Fig. 3.2/d).

The protruding water tank situated on roof tops can be rather unsightly. It can be accommodated in the roof-space if the collector panel is installed in a lower position. Fig. 3.3 shows some possibilities.

For situations where this is not possible, a collector panel has been developed which incorporates the tank in the form of a horizontal cylinder. (Fig. 3.4) The collector panel in this case may be one of the usual types (as described in section 2.4) but a cheaper version may do away with the tubing and provide two continuous layers of water, circulating slowly, over a large cross-sectional area. Due to the large mass of water in the collector this system will have a long response time, but it may be quite suitable for strong radiation climates.

A freedom of positioning the tank in relation to the collector can be obtained by using a small circulating pump. Another advantage of this will be that pipe sizes can be reduced. A pump with a motor around 30 W will be quite sufficient even for large installations.

It is obvious from the expression

$H_g = \Delta t \times C$ ⠀⠀⠀(where C = flow rate $\times$ specific heat of water)

given in section 2.4, that temperature obtained and flow rate are inversely proportionate. Fig. 3.5 shows this relationship. It also shows that with an insufficient flow rate, if the temperature goes beyond the necessary limit, the efficiency of collection is reduced. If a set temperature is required, the flow rate should be adjusted as the radiation input changes.

Thermosyphon collectors are self-regulating. If there is no solar input, there is no heating effect in the absorber, thus there is no force to drive a circulation. The greater the solar input, the faster the flow will be.

In a pumped system a rather sophisticated system of controls would be necessary to operate a modulating valve which could achieve a similar flow rate regulation. It has been suggested that if a D/C motor is used, which responds to variable voltage, and the power is supplied by an array of silicon cells of about 0·25 m² in area, this will be a self-regulating system. With a greater solar intensity a higher voltage is produced, the pump will run faster, the flow rate will be increased, thus the collection temperature will be kept at the desired level.

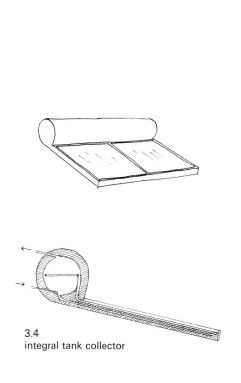

3.4
integral tank collector

3.5
temperature vs flow rate

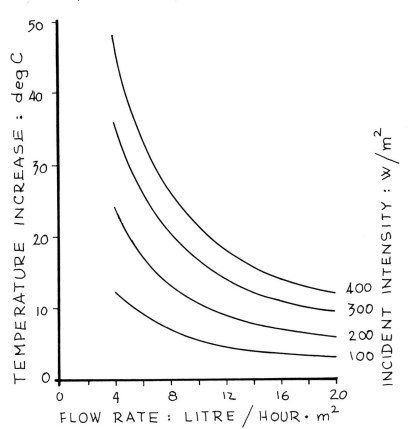

temperature increase, if water entry
temperature = outdoor air temperature
(single glazed collector: $U = 2·7$ W/m²degC)

3.2
Swimming pool heating

The heating of swimming pools requires large quantities of energy. A typical small pool containing 85 m³ water would require almost 100 kWh of heat for every degC increase in temperature. The heat required however, is of a low grade. A comfortable pool temperature is between 21° and 25°C, and it is adequate to supply heat only a few degrees above this temperature. As the efficiency of solar collectors is best with the lowest collection temperature, their use for swimming pool heating is profitable, even in locations where water or space heating would not be economically justified.

The criteria are not very stringent. The success of an installation can best be measured by the extension of the swimming season. Andrassy reported [3] the extension of the normally 50-day swimming season at Princeton (NJ) to 152 days, with water temperature 21°C or more.

In swimming pool installations the circulating pump is normally given, serving the filtration plant. The solar heating panels may be connected in series with this, although some installers prefer the freedom of an independent circulation system, as this enables the setting of the flow rate to the desirable optimum.

The collector used for this purpose is normally of a simpler construction than for domestic hot water heating. As the collection temperature is lower, there is no point installing double glazing. Many ingenious ideas have been tested, such as open flow collectors or the use of plastic hose pipe, tightly fitted into grooves of a metal sheet. The plastic material in this case was mixed with small copper particles to improve its thermal conductivity. Numerous products are commercially available. Some of these are unreasonably expensive, being deliberately aimed at the luxury end of the market.

To get the greatest benefit from such an installation, it is advisable to prevent (or at least reduce) the night-time heat loss from the pool surface by some form of cover. In the simplest case this may be a polythene sheet, rolled up at one end of the pool during the day. Another method uses inflated plastic bags. Floating polystyrene balls have also been used. These can be removed automatically, by lowering a weir, when the flow of the top layer of water into a separate compartment will carry away the light balls.

3.3
Space heating

There are three methods in general use for the utilisation of solar radiation in space heating:

1 the building as collector
This is the common-sense approach in building design. Its basis is a thermally very efficient building, with good insulation placed outside the main mass of wall and roof elements. It has large windows facing the equator, which can be closed by shutters or heavy curtains when there is no solar gain, in order to reduce heat loss. Without such shutters or curtains the glass surface may give an annual cumulative heat loss greater than the annual cumulative solar gain.

2 special building elements
An external enclosing element of the building (a wall or a roof) may be designed in such a way as to act as a collection device.

Eg a massive wall (Fig. 3.6), with the outside painted black, may be covered by one or two sheets of glass. It will act as an absorber, storing some of the heat in its mass and providing an output mechanism through the convection currents induced.

Both methods **1** and **2** can significantly improve the thermal conditions and can substantially reduce the energy requirement of the building, but neither provides a flexibility of controls and neither is precisely predictable in its performance. Given the tolerances and accuracy of building work and the unpredictability of user behaviour, any performance prediction can only provide a rough guidance.

3 flat plate collectors
In this category two basic types must be distinguished:

a water systems
b air systems

The operating principles of water systems are essentially the same as of water heating. The difference is only in magnitude: a single family dwelling may have as much as 80 m² collector area. Due to its size and to the fact that the correspondingly large water tank cannot be positioned above the collector for structural and aesthetic reasons, the circulation must be pumped.

It is usual to provide for the storage of a few days of heat collection, which would bridge over a short period of insufficient collection when the sky is heavily clouded. With water systems the storage medium is generally water, using only the sensible heat capacity of water. (See section 3.7 below.)

Some collectors using air as the collection fluid have been described in section 2.4. These can provide warm air for space heating purposes, although the distribution network as well as the collector itself will, of necessity, be much bulkier. The method does however have certain advantages, eg it avoids the risk of freezing.

3.6
a collector wall

3.4
Water systems

Any solar heating system will consist of five major components:

1 collector
2 storage
3 auxiliary heater
4 distribution system (incl emitters)
5 controls (incl pumps and fans)

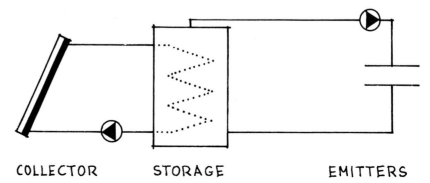

3.7
a solar heating system

COLLECTOR   STORAGE   EMITTERS

The simplest system can be represented as in Fig. 3.7.

The collector may cool below freezing point. Freezing of the water could cause mechanical damage. This can be avoided by three means:

a   keep the circulation going during the frost-danger period (this would obviously mean a substantial heat loss).

b   drain the system. This may be done manually, but rather elaborate automatic systems have also been devised. Fig. 3.8 shows a circuit which would contain

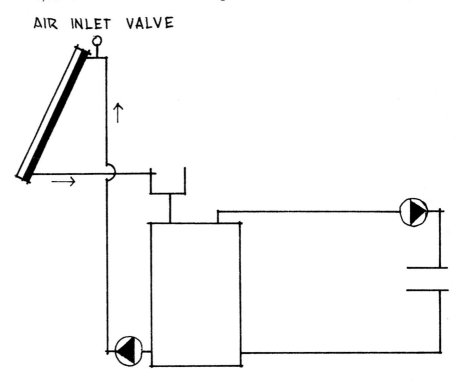

AIR INLET VALVE

3.8
a self-draining circuit

water only when the circulating pump is operating, when there is collection — thus frost danger is avoided. A disadvantage of this method is that the changing water/air content inside the absorber plate may accelerate corrosion.

c   use some anti-freeze agent. Fig. 3.9 shows the extent to which the freezing point is lowered by the addition of various proportions of ethylene glycol. This would slightly reduce the thermal capacity of the liquid (to approx 1·1 Wh/litre degC). The use of anti-freeze agents is only possible with closed circulation systems, ie where there is no consumption of water. If a hot water system is coupled with the heating system, the consumable water must be heated through a heat exchanger.

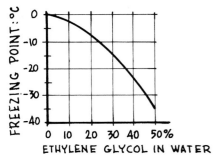

3.9
the effect of anti-freeze

Heat transfer from the collection circuit into the storage may take place by direct flow or through a coil. In the latter case the anti-freeze additive will be needed in the collection circuit only, but the storage temperature will always be 2–3 degC below the collector flow temperature.

The heat emitters (or space heating devices) may be

a   radiator panels
b   ceiling radiators
c   fan convectors
d   embedded floor coils

Radiator panels used in conventional central heating systems are designed to operate with 65°–75°C water. To produce such temperatures by solar collectors would be very inefficient (cf Figs. 2.13 and 2.17) and probably impossible during the winter months. If operated with lower temperatures, greatly increased surface areas would be necessary to get the required heat emission. It may be difficult (as well as expensive) to accommodate such large radiator panels.

Ceiling radiators (either prefabricated panels, or radiant ceilings with embedded coils) may be used. There is no size restriction: the whole of the ceiling may be a radiator, thus it will be possible to run the system on much lower temperatures. The ceiling surface should not be warmer than 32°C, as dictated by reasons of thermal comfort, thus the water flow temperature may be around 35°C.

Fan convector units commercially available, are designed to work on 65°–75°C water, but the heat transfer surface (usually finned tubes) can be increased quite readily, converting these units for using 40°–42°C water.

Floor warming systems using embedded electrical elements are quite popular. Similar systems using hot water pipe coils are also in use. This system would lend itself most readily for use in conjunction with solar heating, as the floor surface temperature being limited to about 25°C, water of around 28°–30°C can be made use of. The thermal capacity of the floor would also assist in the storage of heat and thus in the evening out of weather variations.

A system designed to be fully operative on all days of the year, would be un-economical. In one instance it has been found that if a given area of collector provides all the heat required for 320 days of a year, a doubling of the area will be necessary to cope with a further 30 days, and a second doubling is needed to provide enough heat for the remaining 15 days. The law of diminishing returns prevails. It is more economical to choose the lesser size and rely on an auxiliary heater of some kind for the mid-winter days. This auxiliary heater may be a calorifier (fed by hot water from a boiler), a boiler connected in series or even an electric immersion heater.

Three alternative positions may be considered' as shown in Fig. 3.10:

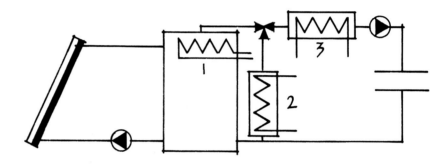

3.10
positions of the auxiliary heater

Position 1 needs the smallest heat output rate, as more time is available for its functioning but it will have to heat a large volume of water unnecessarily and it may prevent collection the following day. It is essentially a slow response system.

Position 2 would make it possible to isolate the heating circuit from the storage and run it as a conventional central heating system, when the storage temperature is below the level required for heating.

Position 3 is the most favoured one, having 'the best of both worlds', having the advantages of (2) but making it possible to utilise the solar collector system as a pre-heater, thus reducing the heat input requirement. Eg if the emitters are designed to use 40°C water (resulting in a return water temperature of, say 25°C) the operation will have three phases (Fig. 3.11):

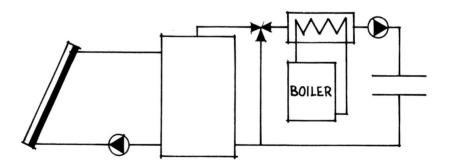

3.11
a three-phase system

a   when collection- (or upper storage) temperature is above 40°C the auxiliary heater does not operate

b   when the collection temperature is between 40° and 28°C, the circulation still flows through the storage and the auxiliary heater works as a topping-up device, bringing the water temperature up to 40°C.

c   when the collection (or storage) temperature drops below the set limit of 28°C, the water flow is short-circuited and the auxiliary heater will work as a conventional central heating system, independently of the solar collector and storage

To allow an optimum collection efficiency under widely varying conditions, a two-stage collection system can be adopted, relying at times on a heat pump system for the transfer and upgrading of the heat (Fig. 3.12).

With high intensity radiation, when a flow temperature greater than (say) 55°C is achieved, the flow is directed into the top tank. Here it will discharge some of its heat to the stored water then the flow enters the lower tank. The coolest part of this is fed back to the collector. With a lesser radiation intensity, thus lower collection temperature, the flow is directed into the lower tank. A compressor type heat pump will use this lower tank as the 'source' and the upper tank as the 'sink'.

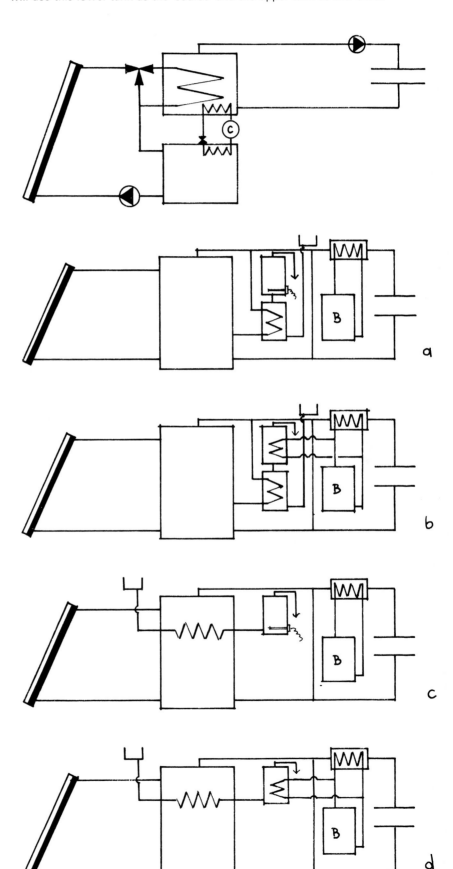

3.12
two-phase collector with heat pump

3.13
combined water and space heating

3.5
Combined water and space heating

Further complications arise if the water heating system is to be linked with this heating system. As the domestic hot water supply should be 60 or 65°C (for kitchen and laundry purposes; for bathroom use 45°C would suffice) there will have to be a topping-up device. If the heating operates at 40°C, the same medium could bring up the cold water to almost the same temperature, but the remaining 20°C must be added by this topping-up device. This could be an independent electric immersion heater or a calorifier fed by the same boiler as the heating. The pre-heating itself can be a flow-through or a storage type system.

Thus we have four possible variants, shown in Fig. 3.13:

a storage type pre-heater electric topping-up

b storage type pre-heater boiler topping-up

c flow-through pre-heater electric topping-up

d flow-through pre-heater boiler topping-up

In the summer when there is no space-heating draw-off, the whole of the storage temperature may increase to above 80°C thus the pre-heater will do all the heating required.

The storage type pre-heaters (**a** and **b** above) have the advantage that more time is available for heating the given volume of water, thus a lesser heat transfer surface will suffice. There is however the need to operate a separate circulating pump. The temperature of water returning into the storage tank from the pre-heating circuit will vary between wide limits. At the beginning of the pre-heating cycle it will approach the cold water temperature (just above 10°C) whilst towards the end of the cycle it will be very near to the flow temperature. This may disturb the stratification pattern in the main storage tank. A system has been devised, using a thermostatically controlled 3-way valve, to discharge into the storage tank near its bottom in the first half of the pre-heating cycle and much further up towards the end of the cycle. (Fig. 3.14)

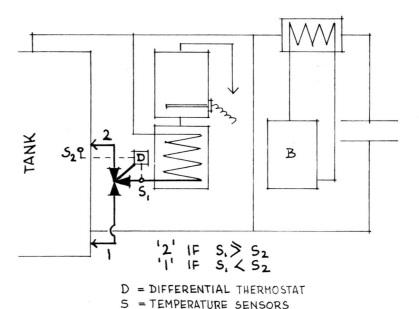

$'2'$ IF $S_1 > S_2$
$'1'$ IF $S_1 < S_2$

D = DIFFERENTIAL THERMOSTAT
S = TEMPERATURE SENSORS

3.14
two-way inlet of the return circuit to assist stratification

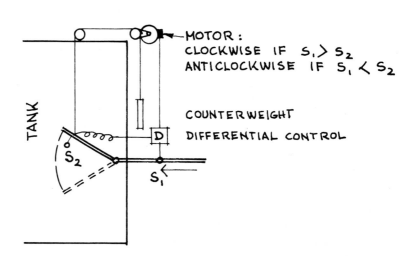

MOTOR:
CLOCKWISE IF $S_1 > S_2$
ANTICLOCKWISE IF $S_1 < S_2$

COUNTERWEIGHT
DIFFERENTIAL CONTROL

3.15
swing-arm tube inlet of the return circuit to assist stratification

There was also a suggestion to use a motorised swing-arm tube (Fig. 3.15) which, controlled by a differential thermostat, would always discharge at a level in the storage where the temperature of the particular layer matches the in-flow temperature. The same system could also be used at the entry of collector flow into the storage tank.

The flow-through type pre-heaters (**c** and **d** above) are much simpler, but must be capable of heating a certain amount of water whilst flowing through. With a large rate of draw-off they may not give as great a heating effect as they should.

A semi-storage type pre-heater is also possible. This consists of a container placed inside the storage tank. The volume of this should correspond to the volume of the maximum demand likely to occur within the period of recovery. (Fig. 3.16)

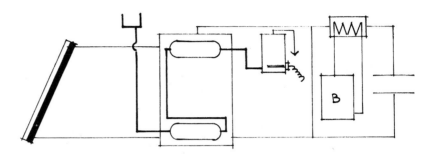

3.16
semi-storage type pre-heater to the hot water system

3.6
Air systems

If warm air is to be used for space heating, it may be worth while considering the use of air as a collection fluid as well. Some suitable collectors have been described in section 2.4. At three points in the system heat must be transferred from solid to air or vice-versa:

a from elements heated by radiation to air in the collector

b from warm air into storage

c from storage to air during recovery periods

There is likely to be a loss in efficiency at each of these points.

The storage must have not only a high thermal capacity but also a large transfer surface. Crushed rock or gravel is often used. The system can take the shape shown in Fig. 3.17. The storage bin may contain fuseable salts in small canisters in lieu of gravel (see section 3.7 below).

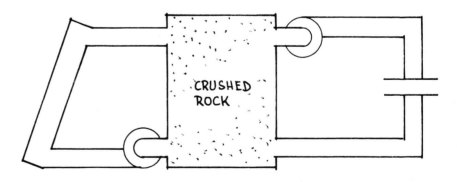

3.17
an air heating system with crushed rock storage

It is difficult to combine this system with a water heating device, but it can be done. The converse, ie waterborne collection and warm air distribution is more practicable. (Fig. 3.18) In the latter case the primary storage is water, but the crushed rock surrounding the water tank acts both as a secondary storage and as a heat transfer device to the warm air heating system. This system can be combined with a water heating device quite readily.

The topping-up device will have to be a furnace of some kind. The water pre-heater may also be a flow-through type device, rather similar to a motorcar radiator used in reverse.

The greatest advantage of the system will appear when the collection is used directly for space heating. An ingenious system has been devised [4] which uses one fan only and facilitates a three stage operation. (Fig. 3.19)

1 collection to space heating
2 collection to storage
3 recovery from storage to space

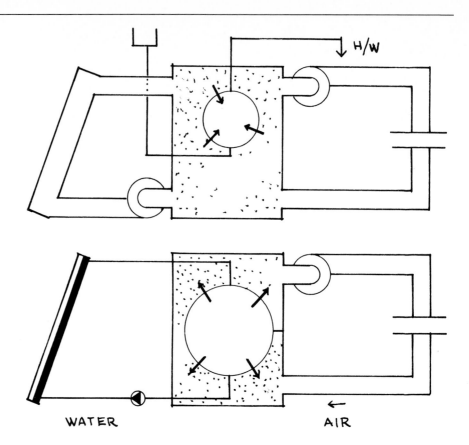

3.18
combined air and water heating (arrows
show direction of heat flow)

WATER   AIR

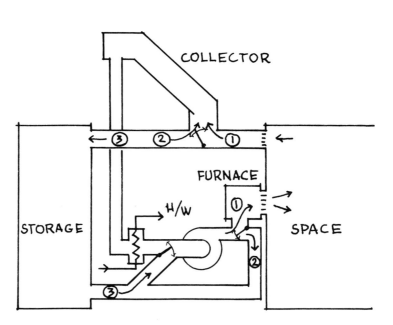

3.19
a single fan air system

The physical arrangement of the operating system within the building and its combination with the building fabric gives rise to an almost infinite number of possibilities.

3.7
Storage of heat

With present day technology inter-seasonal storage of heat (ie storing the summer surplus for use in winter) is economically impracticable. It is however useful and necessary to provide heat storage equal to two or three days' heat requirement (in October or March) to carry the system over one or two days total absence of radiation input.

The most generally used heat stores are based on the sensible heat capacity of materials.

If water is the storage medium, the storage capacity will be the product of the volume, the (volumetric) specific heat of water and the operating temperature range. The volumetric specific heat of water can be taken as 1·16 Wh/litre degC (although it varies slightly with temperature). Thus for example, if the highest temperature obtained is   65°C

and the lowest useful temperature is   30°C

the operating range is   35degC

1 m³ of water would give a storage capacity of

$1000 \times 1.16 \times 35 = 40\ 600$ Wh ($= 40.6$ kWh)

If the daily heat requirement of a house is say 40 kWh and two days' heat is to be stored, the volume required will be

$$\frac{40 \times 2}{40.6} = 1.97 \qquad \text{say 2 m}^3$$

Often the recommended storage volume is related to the collector area. Values in literature range from 50 to 140 litres per m² collector area.

If the storage medium is crushed rock or gravel, the heat storage capacity can be established as the product of the volume of container, the solidity ratio, the density and the specific heat of the stone and the operating temperature range. The solidity ratio may be typically around 0.7 (meaning that there is 30% void between the solids). The density of stones may vary between 2400 and 3000 kg/m³ and their specific heat is around 0.3–0.32 Wh/kg degC.

Thus the 2 m³ store of the above example, if filled with crushed stone in lieu of water, would give a storage capacity of

2 m³ $\times$ 0.7 $\times$ 2800 kg/m³ $\times$ 0.3 Wh/kg degC $\times$ 35 degC =

$= 41\ 160$ Wh ($= 41.16$ kWh)

which is just over half of the storage capacity of an identical volume of water.

The latent heat of crystallisation of some substances may also be utilised for heat storage. The most frequently used such substances are certain low-cost inorganic salt hydrates (eutectic salts), such as sodium sulphate (Glauber's salt). When this is dissolved at a suitable temperature (forming an anhydrous salt solution), a large amount of heat is absorbed from the environment. The process is endothermic. A similar quantity of heat will be released when the solution is cooled and the substance recombines with water, forming salt hydrate crystals in suspension, as crystallisation is an exothermic process.

Table 3.1 gives the transition temperature and the latent heat of reaction for some such salt hydrates. [5]

Table 3.1   Some salts for latent heat storage

	transition temp. °C	latent heat of reaction Wh/kg
$CaCl_2.6H_2O$ calcium chloride	29–39	48
$Na_2CO_3.10H_2O$ sodium carbonate	32–36	74
$Na_2HPO_4.12H_2O$ sodium phosphate	36	73
$Ca(NO_3)_2.4H_2O$ calcium nitrate	40–42	58
$Na_2SO_4–10H_2O$ sodium sulphate	32	67
$Na_2S_2O_3.5H_2O$ sodium thiosulphate	49–51	50

To prevent stratification and segregation, as well as to ensure a large surface of contact even when the substance is molten, it should be enclosed in small sealed containers.

When cooled, these substances may not crystallise, they could suffer 'undercooling'. This may be prevented by using crystallisation catalysts (nucleating agents) eg borax in very small quantities.

The crystallisation velocity of these salts is just below 1 mm/hour degC (difference between solid and liquid). Heat from the container should not be withdrawn at a rate faster than the heat released in crystallisation.

When heated after fully dissolved, the sensible heat capacity of these substances can also be used, which is very similar to that of water.

With ordinary sensible heat storage the temperature is often increased during collection far above the required level. This reduces collection efficiency and increases losses from the store. With latent heat storage the temperature may be kept almost constant, at a level just above the useable temperature. [6]

Unfortunately some of the associated engineering problems (such as the corrosion of containers) have not yet been fully solved.

3.8
Mechanical work

This subject may have a relevance to buildings in only an indirect way, thus only a very brief survey is given. Two basic types of devices can be distinguished:

a expansion engines
b steam engines

Hot air or vapour engines can use temperatures below 100°C. They require a source (heat input) and a sink (heat removal). For small scale installations the Stirling engines seem to be the most suitable. A two cylinder reciprocating engine can reach an efficiency of 30%. The heat input may be provided by hot water from solar collectors (Fig. 3.20), or radiation focused directly on the cylinder head (Fig. 3.21).

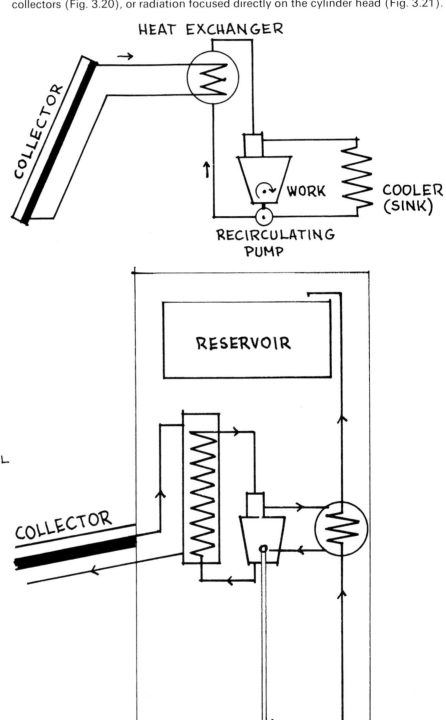

3.20
indirect solar heat engine

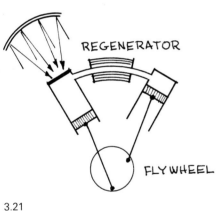

3.21
direct solar heat engine

3.22
pumping with a solar engine

If the work is used for pumping of water the pumped water itself may serve as the heat sink (Fig. 3.22). In this case a cheap semi-concentrating roof trough system has been used as the collector. [8]

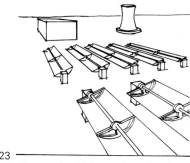

3.23
parabolic troughs driving a steam turbine

3.9
Cooling and
air conditioning

A turbine and compressor combination may be substituted for the two cylinders. Such units can achieve efficiencies in the order of 20%.

Steam may be produced by some form of concentrating device, possibly relying on a flat plate pre-heater. It may be used to drive ordinary reciprocating steam engines. Higher temperature (superheated) steam may be produced by more powerful concentrators and may be used to drive steam turbines.

For large scale installations (in the megawatt order) turbines using approx 650°C steam seem to be the most promising. [7] This temperature can be reached by an array of parabolic trough concentrators, with a concentration rate around 200. (Fig. 3.23)

Cooling by solar energy holds out a great promise, as this is about the only application where the maximum energy demand coincides with the maximum energy collection.

From a surface of high emissivity much heat can be dissipated by radiation to the night sky. By such means a large quantity of water can be cooled overnight. If this is then circulated through the coils of a fan-convector or through coils embedded in the ceiling, a certain degree of space cooling can be achieved. The same system can be used for solar collection and space heating when this is required.

Several examples of such installations are described in part 5.

Refrigeration can be produced by two methods.

a   use of some of the methods described in section 3.8 to produce mechanical work, which in turn will be used to drive a compression type refrigeration cycle. [9] This is the same machine as that described under the title 'heat pump' in section 2.8. (Fig. 3.24)

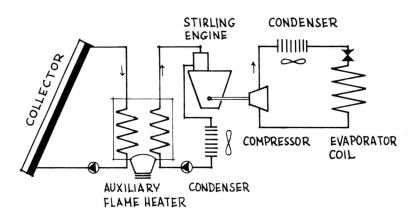

3.24
solar compression refrigerator

b   use of the heat obtained from the sun directly, to drive an absorption refrigerator. This is essentially the same machine as the domestic gas or paraffin (kerosene) operated refrigerator.

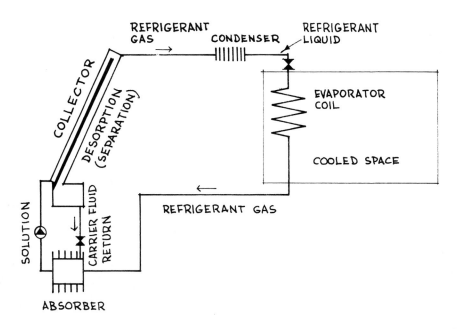

3.25
solar absorption refrigerator

It operates on the following principles:

The *refrigerant* (eg ammonia) is more soluble in the *carrier fluid* (eg water) at low temperatures. Heating by solar radiation (or any other means) will expel some of the refrigerant from the solution. (Fig. 3.25) When this high pressure heated vapour condenses, it will dissipate heat to its environment. The liquid refrigerant will then enter the evaporator through a throttle valve, it will rapidly evaporate, cool, and take up heat from its environment. The warmed refrigerant vapour is re-absorbed in the carrier fluid. This process is also exothermic, thus the absorber will also reject some heat to the environment. This solution will then be pumped back into the solar heater. The pump work input is very small.

Several variants of this absorption cooler are in use or have been tested. One system using water as the refrigerant and lithium bromide as the carrier fluid is commercially available for air conditioning. [10] The lowest temperature it will produce is about +4·5°C, which is not low enough for freezing or refrigerators. Desorption (separation) will take place at around 70°C, thus it is very suitable for use with flat plate collectors of a reasonable quality. The coefficient of performance is about 0·75. Some 5–5·5 m² collector area is necessary to produce 1 kW cooling rate.

Slightly higher coefficients of performance can be achieved with absorbers using a higher desorption temperature, which can be obtained by concentrating devices.

3.10 Distillation

Distillation is a very attractive use of solar energy, as the energy requirement is in the form of low grade heat. The low-technology nature of the system makes it suitable for use in developing countries, where other forms of fresh water supply are non-existent and natural fresh water is scarce.

Non-potable (eg salt or brackish) water is to be evaporated and subsequently condensed. The condensate is distilled water, which is however suitable for consumption, as the condensation took place in the presence of air, thus it contains an adequate amount of dissolved oxygen. Mineral content may be added, if desired, by draining the condensate through eg dolomite gravel.

The simplest system is the hot-box type distiller (Fig. 3.26) In this the black bottom of the tray absorbs the solar radiation, heats the water, which will evaporate. The air-water vapour mixture will develop a convection current. Condensation will occur on the inside of the transparent cover, which is being cooled by the outside ambient air. The condensate will be collected in the channels along the bottom edge of the cover.

This type of distiller may give up to 6 litres of water per m² and day, with an efficiency of up to 70%. (Efficiency is defined as the ratio of the heat used to heat and evaporate the water in the tray, to the incident radiant energy.) [11]

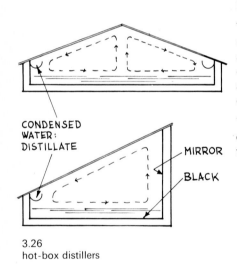

3.26
hot-box distillers

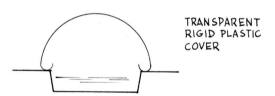

TRANSPARENT
RIGID PLASTIC
COVER

TRENCH LINED WITH BLACK PLASTIC FILM

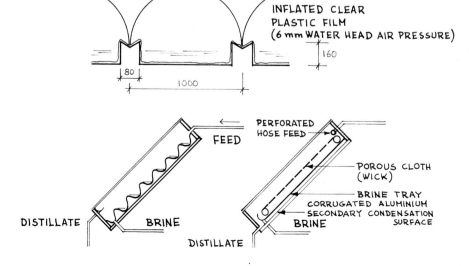

INFLATED CLEAR
PLASTIC FILM
(6 mm WATER HEAD AIR PRESSURE)

3.27
various solar distillers

Numerous variants of this system have been developed and many large scale installations are in operation. (Fig. 3.27)

More elaborate installations can achieve slightly higher efficiencies, but the capital cost is greater. (Fig. 3.28)

Suggestions have been made and experiments have been carried out in Montreal, to include a distiller as the roof of a small prefabricated bathroom unit, suitable for low-cost housing in developing countries. It includes a rainwater collection device as well as a water recycling system. (Fig. 3.29)

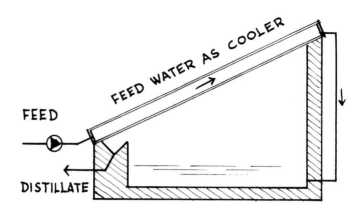

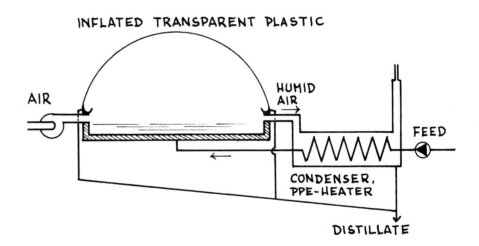

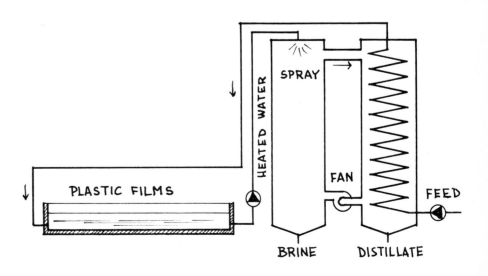

3.28
mechanised distillers

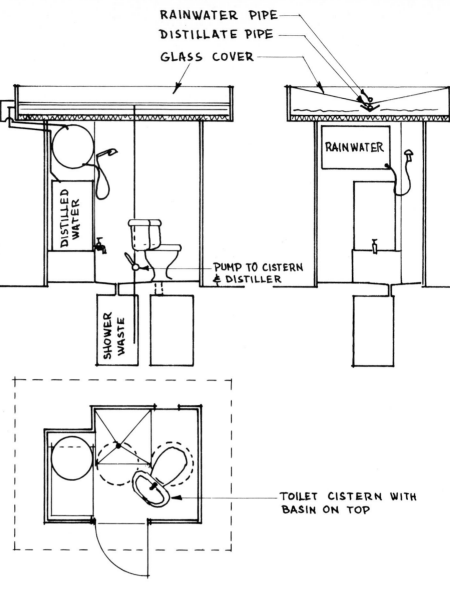

RAINWATER PIPE
DISTILLATE PIPE
GLASS COVER

DISTILLED WATER

RAINWATER

PUMP TO CISTERN & DISTILLER

SHOWER WASTE

TOILET CISTERN WITH BASIN ON TOP

3.29
an integrated distiller-roof

3.11
Drying

Drying of various crops and products (grain, fruit, eg sultanas, palm or sugar cane juice, rubber, meat, timber) was traditionally done under natural conditions, perhaps only providing cover as a protection against rain. This, during the present century, has largely been replaced by various machines using electricity, oil or some other fuel, achieving a controllable performance and an accelerated process. However, the traditional source of heat, the sun's radiation, can also be harnessed and controlled.

Drying can be carried out two ways:

a   the product itself may be exposed to the sun in a covered tray of some kind, or

b   air may be heated in some device and blown through or over the product to be dried.

FOLD-OUT METAL REFLECTORS

3.30
a portable drier

The simplest dryer is a shallow tilted tray with two fold-out mirrors (cf concentrating devices, section 2.10 and Fig. 2.27) giving a concentration rate of 2. (Fig. 3.30) This is a small, portable device. Larger, permanent installations are more likely to use the second method. Of the three examples shown in Fig. 3.31 the first two pass an air current behind an opaque surface heated by the sun, whilst the last one passes air through both sides of the absorber plate, which is under a transparent cover.

With the shortage of oil the mining of oil shales is becoming increasingly important. These usually contain 30–35% of water. Mechanical drying would use a significant part of the oil produced. The substance lends itself to drying in collector-trays rather than blowing solar heated air through it.

3.12
Cooking

In the preparation of food three kinds of processes can be distinguished:

a   direct fire   —up to 1000°C
b   oven cooking —200°–250°C
c   boiling   —100°C

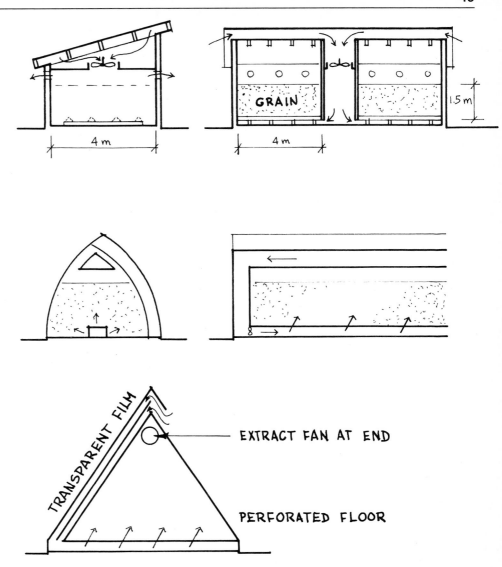

EXTRACT FAN AT END

TRANSPARENT FILM

PERFORATED FLOOR

GRAIN

1.5 m

4 m

4 m

3.31
drier buildings

The food itself will not be heated above 100°C in any case, as its water content sets a limit to the temperature increase, but in most processes the heat must be supplied at a substantially higher temperature.

The average gas ring or electric hotplate delivers heat at a rate around 1 kW, which is enough to boil 2 litres of water in about 10 minutes. To match this rate, some 2 m² collector area will be required. [12]

Two basic types of solar cookers have evolved:

1   the direct, or focusing cooker, where the foodstuff (or the pot containing it) is placed at the focus of a parabolic mirror. A concentration rate between 20 and 100 will give a performance similar to that of an open fire.

2   the box or oven type cooker, which is an insulated chamber with a window on one side, through which radiation will enter, possibly from a number of plane mirrors or a parabolic reflector. A concentration rate between 2 and 4 will give about the same performance as the average domestic oven.

Much effort has been spent on devising the cheapest possible cooker, which will be suitable for use in developing countries and can be produced by semi-skilled labour without heavy machinery.

Fig. 3.32/a shows an early, pressed metal parabolic reflector cooker, b and c are moulded plastic reflectors whilst d and e are collapsible, umbrella type reflectors made of a metallised plastic film laminated to cloth.

These cookers would obviously work only when there is enough direct solar radiation. The box or oven type cookers may have at least some short term storage facility. The cooker devised by Maria Telkes is a basket with heavy clay lining, a glazed cover and four fold-out mirrors. The rounded bottom facilitates a tilting adjustment when sitting in a shallow hole in the ground. [13] Fig. 3.33 shows this basket-oven as well as a slightly more elaborate version having an aluminium body and a tubular frame.

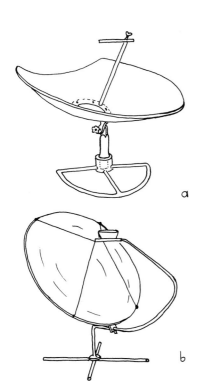

a

b

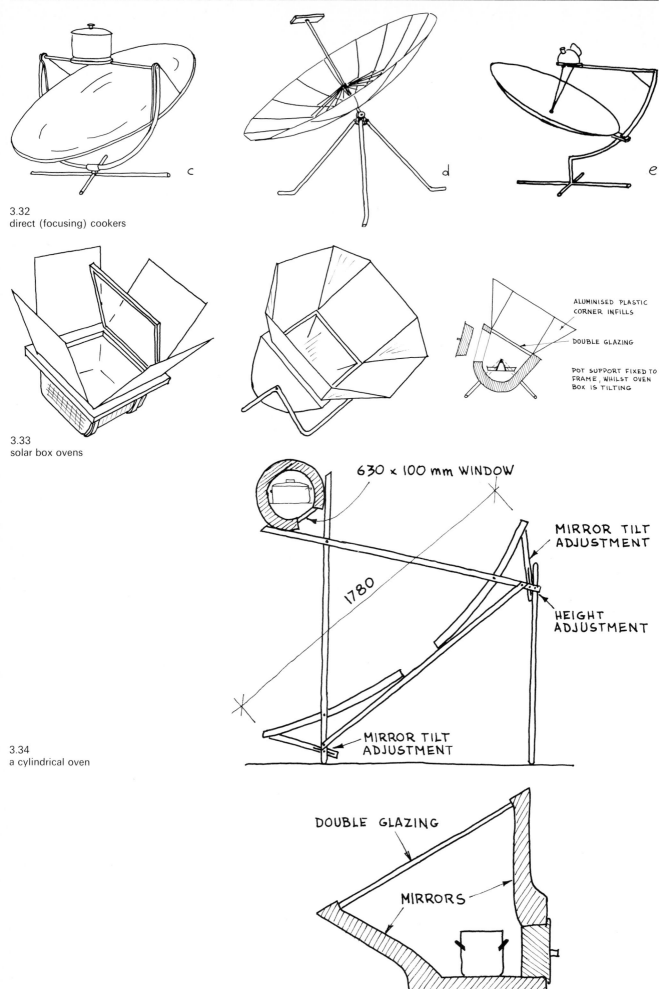

3.32
direct (focusing) cookers

3.33
solar box ovens

ALUMINISED PLASTIC
CORNER INFILLS

DOUBLE GLAZING

POT SUPPORT FIXED TO
FRAME, WHILST OVEN
BOX IS TILTING

630 × 100 mm WINDOW

MIRROR TILT
ADJUSTMENT

1780

HEIGHT
ADJUSTMENT

MIRROR TILT
ADJUSTMENT

3.34
a cylindrical oven

DOUBLE GLAZING

MIRRORS

3.35
an internally mirrored oven

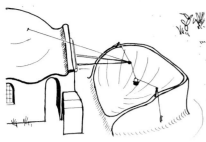

3.36
a fixed reflector cooker

3.13
Electricity production

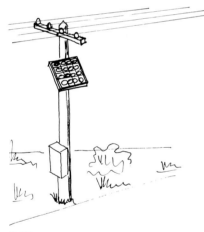

3.37
small photoelectric arrays

A cylindrical oven, using a pair of cylindro-parabolic mirrors has been developed in Portugal (Fig. 3.34) and another version, where the mirrors are inside the glass cover of the oven, was produced in Cairo. (Fig. 3.35)

All these cookers can achieve temperatures between 180° and 250°C, more than adequate for baking. The direct cookers are more suitable for boiling.

An interesting idea has been developed by the Dutch, as shown in Fig. 3.36. Here a spherical reflector is formed in mud, lined with aluminium foil applied with a bitumenous adhesive. As the reflector is fixed, the focal point moves inversely with the sun's movement. The cooking pot is suspended on a chain and can be moved by horizontal ropes to follow the movement of the focal point.

Electricity can be produced from solar radiation by two methods:

a   direct conversion by photoelectric or thermoelectric processes
b   by producing mechanical work which will then be used to drive conventional electric generators.

The principles of direct conversion to electricity have been discussed in section 2.2. Practical applications to date (apart from space applications) are restricted to small scale power generation in locations where no other form of electricity supply is available. Various systems of photoelectric cell arrays are in use for navigation aids, for telephone amplifiers in remote areas or as battery chargers for boats and caravans. (Fig. 3.37) Polgar reported the setting up of an educational television network in Africa, by French television, where village school TV receivers are powered by photoelectric cell arrays.

These are certainly competitive, where the only alternative would be the use of disposable dry-cell batteries, but more expensive than any other method of electricity generation. In the last year or two the cost of cells has been reduced from £3000 to about £300 per m². It is expected that the commercial development of grown ribbon cells will reduce this cost to about £30/m² within a few years.

With the recent drastic increases in electricity charges a cost of £15/m² would already be competitive, if the other great problem, ie the storage of electricity could be solved. Batteries are expensive and bulky. This limits the use of solar electricity to uses which occur during sunshine periods or to only small scale uses outside such periods.

With large scale production of mechanical work for the purposes of electricity generation the main difficulty seems to be the transfer of energy from the very extensive catchment area to the turbines. [7] (An area of 4 hectares would produce 800 kW for 6 hours a day on annual average.) This transfer of energy can take two forms:

optical
thermal

Optical transfer would mean the use of mirrors, ie the transfer of radiant energy before it is converted to heat. A very high rate of concentration would be produced on a central 'boiler', which would be closely linked to the turbines. This boiler would probably be located on a high tower to avoid very low angles of incidence. (Fig. 3.38/a)

Thermal transfer may mean the generation of heat in many small concentrators located over a large area and the transfer of steam through well insulated pipes to the turbine (Fig. 3.38/b). Helium gas has been tested as a transfer fluid, but was found to require a large power for the circulating fans. The vapour of a sodium-potassium alloy (NaK) has been used successfully to transport heat at 700°C. The 'heat pipe' principle could be put to use to solve this transport problem. (Fig. 3.39)

This is essentially a sealed tube, containing a small quantity of a suitable thermal fluid (eg water, Dow-therm, potassium), which changes phase at the appropriate operating temperature. Over most of the length of the pipe, as it is heated, the liquid evaporates and rises to the higher end, transporting heat equal to its latent heat of evaporation. Here it condenses, releasing its heat. The liquid flows back (either by gravity or by capillary action) and the cycle is repeated. The process is very fast. Such a pipe can be a thousand times better conductor than a solid copper rod of the same cross section.

The storage problem also arises with the solar thermo-mechanical electricity generation systems. Large scale installations at the present are only feasible if coupled with hydro-electric plants. The water turbines can be closed down fairly readily when the solar supply can meet the demand. Any surplus solar electricity

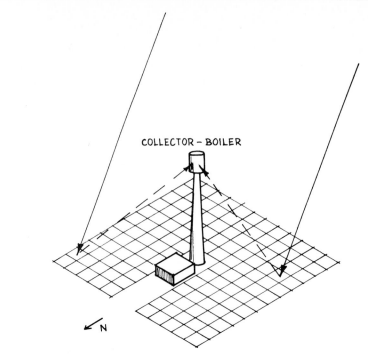

a: A FIELD OF ADJUSTABLE PLANE MIRRORS

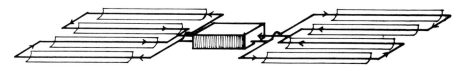

b: A FIELD OF TILTING PARABOLIC TROUGHS

3.38
schemes for power generation

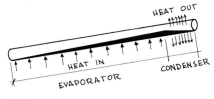

3.39
the heat pipe

3.14
Special uses

can be stored by using it to pump water back into the higher level reservoir.

On the longer term and with general validity solar electricity production will become feasible if reversible chemical reactions become available on a large scale to store the energy easily, in large quantities and inexpensively. The splitting of water by electrolysis into hydrogen and oxygen and thus the production of hydrogen fuel seems to be one of the most promising methods.

There are certain applications where radiant solar heat cannot be replaced by any other form of heat, regardless of its high cost. The various uses of solar furnaces fall into this category. The 'pure' form of heat obtainable even in vacuum chambers gives a unique opportunity for research in various areas of physics and chemistry. High temperature processing of certain materials, such as rare earths, ultra-refractory oxides, or metals such as tungsten is already carried out in solar furnaces on an industrial scale. Solar furnaces also provide a unique opportunity for thermal shock studies.

Laszlo [14] lists a number of special operations for which solar furnaces provide an opportunity:

— growing of single crystals
— zone refining
— high temperature fabrication of thermocouples, and ceramics; welding of ceramics to metals, cutting and casting of refractory materials

References

1 Chinnery, D N W
Solar water heating in South Africa
Nat Bldg Res Inst. Bulletin 44

2 Commonwealth Scientific & Industrial Research Organisation (Australia)
Solar water heaters
Div Mech Eng Circular No. 2
CSIRO 1964

3 Andrassy, Stella
Solar water heaters
paper S 96
'New Sources of Energy' Conf. Rome, 1961
UN 1964

4 Löf, G O G et al.
Design and performance of a domestic heating system employing solar heated air
paper S 114
ibid

5 Telkes, Maria
A review of solar house heating
in Heating and Ventilating, Sept. 1949

6 Telkes, Maria
Performance of the solar heated house at Dover, Mass.
'Space heating with solar energy'
course-symposium at MIT, Aug. 1950

7 Oman, H & Bishop, C J
Feasibility of solar power for Seattle, Washington
paper E 3
'The sun in the service of mankind' Conf. Paris, 1973

8 Alexandroff, Guennec & Girardier
The use of solar energy for water pumping in arid areas
Uni of Dakar, 1973

9 Teagan, W P & Sargent, S L
A solar powered heating/cooling system
paper E 94
'The sun in the service of mankind' Conf. Paris, 1973

10 Chinappa, J C W
. . . solar operated multi-stage vapour absorption air conditioners . . .
paper E 97
ibid

11 Gomella, C
Use of solar energy for the production of fresh water
summary GR 19
'New Sources of Energy' Conf. Rome, 1961
UN 1964

12 Löf G O G
Solar cooking
summary GR 16
ibid

13 UN—FAO—Nutrition Division
. . . the Telkes solar oven and the Wisconsin solar stove
paper S 116
ibid

14 Laszlo, T S
New techniques and possibilities in solar furnaces
paper S 5
ibid

Part 4 Sun and Building

4.1
Solar heat gain

Solar radiation affects buildings in two ways:

1 entering through windows, absorbed by surfaces inside the building, thus causing a heating effect

2 absorbed by outside surfaces of the building, creating a heat input into the fabric, which will be partly emitted to the outside, mostly by convection, but will partly be conducted through the fabric and subsequently emitted to the inside

Both effects can be influenced (if not determined) by the designer. The transmission through windows is determined by

a orientation of the window (thus the intensity of radiation incident on its surface)

b size of the window

c type of glazing (clear, heat absorbing, heat rejecting or photochromatic glasses)

d shading devices, either external (grilles, louvres, canopies, awnings, shutters) or internal (blinds, curtains)

If the intensity of radiation (I) incident on the window is known, the rate of heat gain will be

$Q_s = I \ A \ \tau$ where Q_s = solar heat gain (W)
A = window area (m²)
τ = solar gain factor

the latter includes the transmission coefficient (θ) plus an experimentally determined proportion of the absorption coefficient, corresponding to the proportion of the heat absorbed by the glass that will subsequently be emitted inwards (see Fig. 4.12).

In addition to the solar gain there will be a conduction heat flow through the window

$Q_c = A \ U \ \Delta t$ where U = transmittance (W/m² degC)
Δt = temperature difference (degC) ($t_o - t_i$ for heat gain)

Q_s and Q_c may have opposite signs, eg on a sunny day in winter, when there may be a simultaneous solar gain and conduction heat loss.

The same equation is used to determine heat flow rate through solid walls or roofs. If there is solar radiation incident on opaque surfaces, its heating effect (over and above the effect of air temperature) can be expressed as the 'sol-air excess' temperature

$$t_e = \frac{1 \times a}{f}$$

where a = absorption coefficient
f = surface (film) conductance (W/m² degC)

It may be noted that this is the same expression as that given in section 2.3 for the equilibrium temperature of irradiated surfaces.

The *sol-air temperature* is the sum of this and the outdoor air temperature

$$t_s = t_0 + t_e = t_0 + \frac{1 \times a}{f}$$

The heat flow due to the combined effect of warm air and solar radiation can be found if for Δt we use

$t_s - t_i$ in lieu of the $t_0 - t_i$

It will be seen from this expression that the rate of heat flow through solid walls and roofs is determined by

a   orientation of the surface (thus the intensity of radiation incident on it)

b   the area of the exposed surface

c   the absorption coefficient of the surface

d   the surface (film) conductance, governing the heat emission, depending partly on surface qualities (texture, colour) but more on the velocity of air movement passing the surface, thus on the degree of exposure.

The surface or film conductance is a composite quantity, allowing for radiant as well as convective heat transfer. Its value for building surfaces is between 8 and 80 W/m² degC, normally taken between 12 and 20 W/m² degC. In actual fact its value varies with air velocity, as shown in Table 4.1.

Table 4.1   Film conductances

	1	4	7	10 m/s
plaster, textured render	20	36	52	68 W/m² degC
brick, cement render	16	29	42	55
glass, white paint	9	17	25	33

Absorption and emission coefficients of a surface are the same for a particular wavelength of radiation, but not necessarily so for different wavelengths (cf selective surfaces, in section 2.4). This can be made use of when considering building surfaces exposed to radiation. The bulk of solar radiation is between 0·3 and 3 μm, whereas building surfaces in the 10°–30°C range emit predominantly in the 6–30 μm band (see Fig. 2.8).

Table 4.2 below gives absorption coefficients for solar radiation and emission coefficients for long-wave infra-red (average values).

Table 4.2   Absorption and emission coefficients

	a (solar)	e (low temp.)
black, non-metallic surfaces	0·92	0·94
red brick, stone, tile	0·73	0·90
yellow and buff brick, stone	0·60	0·90
cream brick, tile, plaster	0·40	0·50
window glass (clear)	transparent	0·92
dull glass, aluminium, galv. steel	0·50	0·25
polished brass, copper	0·40	0·03
polished aluminium, chrome	0·20	0·10
white paint	0·20	0·80

It will be seen from this that for example a white paint and a bright metal surface may have the same absorption coefficient (0·2) but the former will have an emission coefficient of 0·8 and the latter only 0·1. Thus both may absorb the same amount of heat, but the white will be cooler, as it emits more.

4.2 Desirability of solar heat gain

The relationship of sun and building can be examined from two points of view:

1   exclusion of solar radiation, as it would cause overheating, an extra load on air conditioning, glare problems or deterioration of materials

2   ensuring adequate sunlight, either to obtain heat when it is in short supply or purely for its psychological effect

No doubt that in tropical climates the first of the two attitudes will dominate [2], whilst in cold-winter regions the latter will prevail. It has however been shown that even in moderate climates severe overheating problems can occur [1]. The all-glass wall has been proved to be thermally inferior to the solid wall having small windows, as it would give a larger heat loss in winter and an unnecessary solar heat gain in the summer.

Others have advocated the use of extensive glazing on south facing walls. Several buildings having such glazing, both in the US and in the UK are described as 'solar houses'.

This apparent contradiction can be resolved in terms of time: both statements may be true at different times of the year.

It has been reported (see section 5.4) that in one particular solar house 33% of the heat requirement is gained through the windows, 49% by the collector system and only 18% by an auxiliary heat source. Another building (see section 5.17) claims to be heated exclusively by the solar heat gain through the windows.

Solar heat gain through windows, even in winter, is common experience and its magnitude can be verified by a simple calculation.

If indoor temperature is $t_i = 20°C$
outdoor temperature is $t_o = 0°C$
thermal transmittance taken as $U = 5$ W/m² degC
radiation intensity is $I = 400$ W/m²
solar gain factor is $\tau = 0.8$

for 1 m² we have

$Q_{loss} = 5 \times 20 = 100$ W/m²

$Q_{gain} = 400 \times 0.8 = 320$ W/m²

The gain is more than three times as much as the loss. However, the above is true only for the duration of sunshine. Even if we take a sunny day in winter, for the 24 hour period the total loss may be greater than the total gain.

At the very best in December–January a south facing window (in London) may get just over 2000 Wh/m² radiation in a day. The heat loss during the same 24 hours may be $100 \times 24 = 2400$ Wh/m². And most days will have much less radiation. Double glazing would improve the comparison, but clearly, what is needed is some form of control—control of solar gain, but also co-ordinated control of all thermal factors of the building.

4.3 Solar geometry

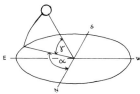

4.1
altitude (γ) and (α) azimuth angles

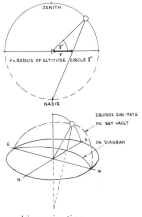

4.2
stereographic projection

* In [2] a set of stereographic sun-path diagrams for latitudes 0–44° N and S are included. [4] gives similar diagrams for UK latitudes.

The most efficient solar control is provided by external shading devices. These can be designed only if the position of the sun relative to the building face is known, ie if the solar geometry is determined.

As mentioned in section 1.9, the position of the sun at any point in time is defined in terms of two angles: altitude (γ) and azimuth (α), as shown in Fig. 4.1.

These angles can be found for any hour of the year from almanacs or from sun-path diagrams of various kinds. By far the best known of these are the stereographic solar charts. The method of projection is explained in Fig. 4.2 and an example is shown in Fig. 4.3.* The two angles can be read directly:

α, or the azimuth angle by projecting the time point from the centre to the perimeter scale

γ, or altitude angle along the set of concentric circles

If the sun's position is to be related to the building, or rather to a particular vertical wall of the building, the *shadow angle protractor* will have to be used. This will give readings of two further angles:

δ, or horizontal shadow angle, the azimuth difference (if orientation, ie the wall azimuth is ω, then $\delta = \omega - \alpha$)

ε, or vertical shadow angle, ie the altitude angle of the sun, projected parallel to the wall on to a vertical plane which is perpendicular to the wall. This will normally be the plane of a section of the building. When δ is zero, ie the sun is directly opposite the wall, $\varepsilon = \gamma$. In all other cases $\varepsilon > \gamma$. The relationship can be expressed as $\tan \varepsilon = \sec \delta \times \tan \gamma$

Fig. 4.4 shows the shadow angle protractor and Fig. 4.5 illustrates the method of establishing shadow angles.

The shadow angles thus obtained can be used to predict the performance of a shading device or the penetration of the sun at any given time, or the overshadowing effect of other buildings or objects. The method is however not only a checking tool, it can be used as a design tool in a more direct way.

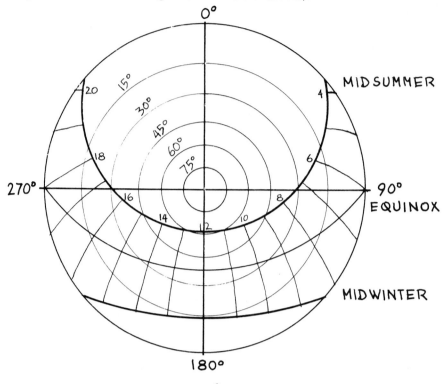

4.3
a sun-path diagram

4.4
the shadow angle contractor

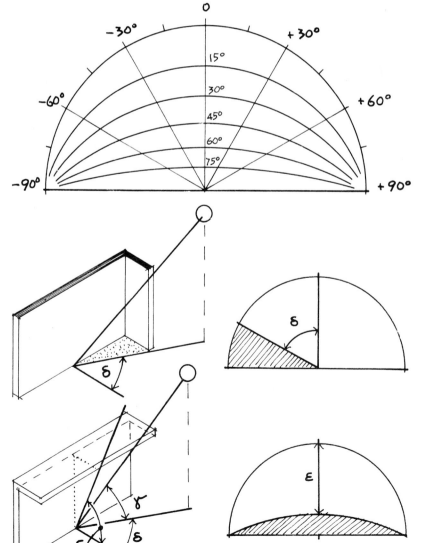

4.5
establishing shadow angles

4.4
Shading devices

The performance of a shading device can be defined in terms of a *shading mask*. Three basic types of devices can be distinguished. [6]

A vertical device (Fig. 4.6) will be characterised by a horizontal shadow angle and will give a sector shaped shading mask.

A horizontal device (Fig. 4.7) will be characterised by a vertical shadow angle and will give a segmental shading mask.

An egg-crate type system, ie a combination of vertical and horizontal elements (Fig. 4.8), will give a combined mask, characterised by both vertical and horizontal shadow angles.

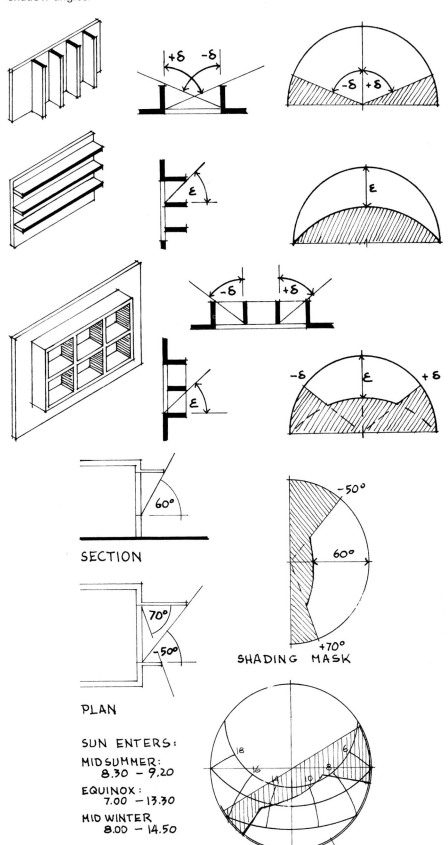

4.6
a vertical shading device

4.7
a horizontal shading device

4.8
an egg-crate type device

SECTION

PLAN

SUN ENTERS:

MIDSUMMER:
8.30 – 9.20

EQUINOX:
7.00 – 13.30

MID WINTER
8.00 – 14.50

SHADING MASK

4.9
a method of determining the period of shading

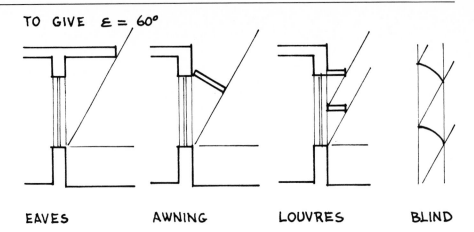

TO GIVE $\varepsilon = 60°$

4.10
various devices for a set performance

EAVES   **AWNING**   **LOUVRES**   **BLIND**

On the basis of plans and sections the shading mask of any device can be constructed by using the protractor. When this shading mask is laid over the sun-path diagram, the period of shading can be read immediately along the date and hour lines. (Fig. 4.9)

The size and physical make-up of devices does not matter from the point of view of geometry. Fig. 4.10 shows that many different devices can have the same performance. Thus the designer may decide early in the process the required shading performance, ie the shading mask, still preserving his freedom for the selection of the actual device.

4.5
Solar control

Fixed shading devices are purely negative controls, ie they exclude the sunshine. Adjustable devices (louvres, bris-soleils) are possible, but rather expensive. However, even fixed devices can be designed to give a selective performance, ie to admit the sun when it is desirable and exclude it when it would cause overheating. The overheated period can be outlined on the sun-path diagram. Then a shading mask should be constructed to match the shape of the overheated period as closely as practicable.

Fig. 4.11 shows that a horizontal device, eg wide overhanging eaves over a south

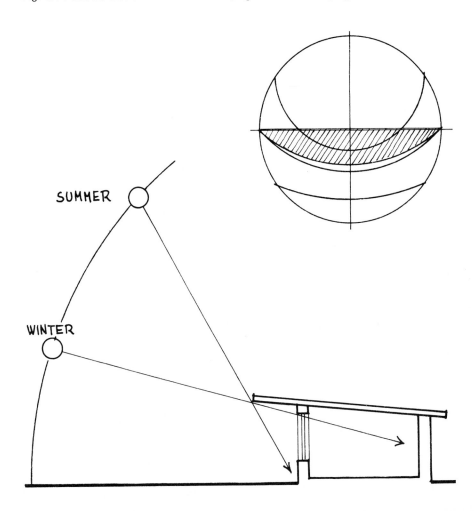

SUMMER

WINTER

4.11
summer/winter shading, admitting sun in winter but excluding it in summer

facing window would exclude the high altitude summer sun but admit the radiation in winter when the sun is at a low angle. It is also shown how this is reflected by the shading mask and sun-path diagram. There is a close match between the shape of the overheated period and the shading mask.

Special glasses can also be used for solar control. The heat absorbing glasses have selective absorption properties, whilst the heat rejecting glasses have a selective reflectance. Fig. 4.12 gives spectral transmission characteristics of some glasses and diagrammatic representation of the reflection/transmission/absorption/re-emission

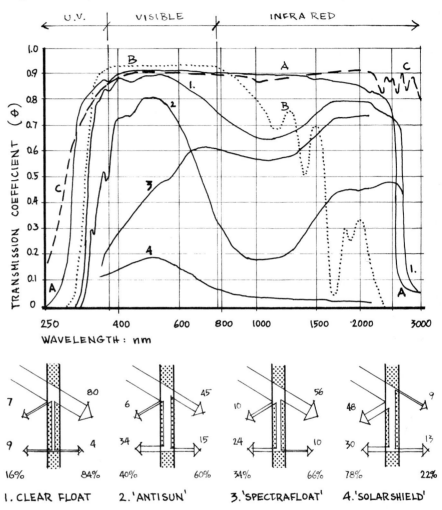

4.12
spectral transmission characteristics of various glazing materials

A. Glass with 0·02 Fe_2O_3 content, 6 mm
B. Polymethyl methacrylate sheet, 6 mm
C. P.V.C. film, 0·035 mm

processes. A curve for an ordinary clear float glass is included for comparison and the transmission curve for an iron-free glass shows the maximum transmission that can be obtained from any glass. Transmission curves for some transparent plastic materials are also shown for comparative purposes.

These special glasses will reduce the radiant heat transmission, but once installed, they will act as controls all the time, do not distinguish between summer and winter conditions.

4.6
Thermal controls

The concept of the building as a 'climatic filter' has been put forward by several authors. To carry the simile further: it should be a selective filter, admitting environmental influences which are desirable and excluding the undesirable ones.

Solar controls, both shading devices and special glasses, may be considered as such selective filters. Their performance however cannot be considered in isolation. They will thermally interact with the whole of the building and its use functions. The thermal behaviour of the building will be determined by the above discussed factors:

window sizes and orientations
type of glazing and any shading device
surface qualities, size and exposure of solid elements

but also by

thermal insulation of enclosing elements
thermal capacity of the building fabric
the relative position of insulation and capacity
ventilation and its variability

All these factors must be considered in relation to the use of the building, heat output of lighting, of persons and processes and the periodicity of these. To use Professor Page's analogy: [9] the designer must be a conductor, co-ordinating the orchestration, the performance of a multitude of instruments.

In some situations the means listed above, ie the *passive thermal controls* may achieve satisfactory indoor conditions. But even if comfort cannot be ensured by such means alone, the good design will greatly reduce the task of *active controls*, ie of installations using some form of energy input, such as heating or air conditioning.

4.7
Thermal capacity

In moderate climates, such as the UK, the extreme swing of temperatures between summer and winter may not be very great, but the short term rate of change may be as large as in the most extreme climates. It is such short term changes that can be best buffered or evened out by a large thermal mass.

In mid-winter for example, on a clear day there may be a large solar input, which may cause overheating. Blinds are drawn, or windows are opened, the heat is rejected or wasted. If this heat could be stored in a massive building fabric, it would be preserved for use during the cold night. (Clear skies in winter mean generally the coldest night but the best solar gain during the day.) If the insulation is on the inside of the thermal mass, much heat will be absorbed in the mass during the day but most of this stored heat will be dissipated to the outside during the night. If we have the thermal capacity on the inside of the insulation, much better results will be achieved, although the amount of heat entering the storage during the sunshine period is reduced.

It would be ideal to have some insulating outer cover which is removable or open-able, thus it could admit solar heat into the massive wall during the solar gain period, but which could be closed after sunset, to preserve the stored heat and allow it to be dissipated to the indoors. Alternatively a system could be devised which transfers the heat from the outer surface to the thermal mass, located inside of a substantial insulation. Overnight the transfer system could be closed down.

This is in fact the principle behind all solar heating systems. The collector is on the outer surface, which will collect the solar gain and transfer it to storage, from where it can be drawn off on demand. The simple systems are less readily controllable. A desire for flexibility of control can lead to great complexities. It is only the details of the solution that varies and leads to the great diversity of solar houses.

4.8
Thermal insulation

Whether solar houses or ordinary buildings, the thermal insulating properties of the enclosing elements are equally important. A study carried out by the National Building Agency [3] in 1969 proved that doubling of our insulation standards would pay for itself in 4 years in the south of England and in 2 years in northern Scotland. (!) Some research workers have suggested that greater benefits and greater energy economies can be achieved by improving the thermal insulation standards than by any solar heating installation.

Neither of these views is disputable. It is however strongly suggested that we can go further than reducing the heat demand by insulation. Important as it is, the use of insulation is not enough. It must be used intelligently, in the correct position in relation to the thermal mass.

There is no point in installing an expensive solar heating system if the building is thermally inefficient. If the building is well insulated and thermally well designed, the contribution of a solar heating system will be proportionately greater and may be able to supply almost all the (reduced) energy requirements.

Present building regulations [5] prescribe the following minima of insulation (ie the maximum acceptable U-values):

roof (incl ceiling) 1·42 W/m² degC
wall (without window) 1·70

In strict economic terms (late 1973 fuel prices) the absolute minimum insulation (highest U-values) should be:

roof (incl ceiling) 0·57 W/m² degC
wall (without window) 0·85
wall with window (average) 1·20

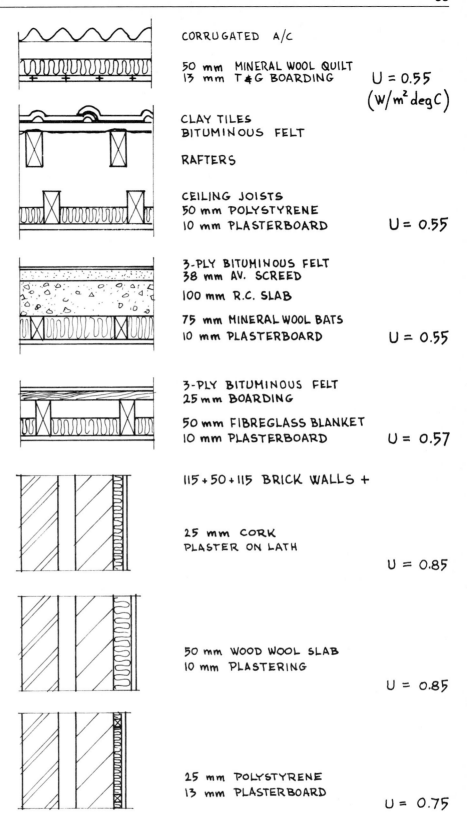

CORRUGATED A/C

50 mm MINERAL WOOL QUILT
13 mm T&G BOARDING

$U = 0.55$
$(W/m^2 \text{deg} C)$

CLAY TILES
BITUMINOUS FELT

RAFTERS

CEILING JOISTS
50 mm POLYSTYRENE
10 mm PLASTERBOARD

$U = 0.55$

3-PLY BITUMINOUS FELT
38 mm AV. SCREED

100 mm R.C. SLAB

75 mm MINERAL WOOL BATS
10 mm PLASTERBOARD

$U = 0.55$

3-PLY BITUMINOUS FELT
25 mm BOARDING

50 mm FIBREGLASS BLANKET
10 mm PLASTERBOARD

$U = 0.57$

115 + 50 + 115 BRICK WALLS +

25 mm CORK
PLASTER ON LATH

$U = 0.85$

50 mm WOOD WOOL SLAB
10 mm PLASTERING

$U = 0.85$

25 mm POLYSTYRENE
13 mm PLASTERBOARD

$U = 0.75$

4.13
some constructions with satisfactory
U-values

Fig. 4.13 shows some constructional examples which would satisfy these requirements.

Some of the buildings described in part 5 have walls with U-values as low as 0·3 W/m² degC.

The heat lost from the building by conduction through the enclosing elements will be

$Q_c = (\Sigma\, A\, U)\Delta t$ where A = area of each element (m²)
U = thermal transmittance of each element
(W/m² degC)
Δt = temperature difference (degC)

The cost of insulating materials is not necessarily proportionate to their insulating value. Their cost-effectiveness can be expressed by using the 'cost-index' concept. This index can be computed the following way:

1 take the cost (in pence) per m² for supply and fixing of the specified thickness of the material

2 calculate the conductance of the specified thickness, from the k-value of the material

3 the product of this cost and conductance* is the cost index, which represents the cost (in pence) of the insulation which would give a conductance of 1 W/m² degC

Thus the smaller the cost index, the better the cost-effectiveness. Table 4.3 gives the cost indices of some frequently used insulating materials. Generally the lower cost index materials perform only an insulating function. Materials which give also a finish or perform also some structural function, usually have a higher cost index.

Table 4.3   Cost indices of insulating materials

Walls	k	75 mm	100 mm	150 mm
1 solid clinker block	0·331	587	487	440
2 hollow clay blocks	0·259	473	414	436
3 l.w. aggregate block	0·209	418	382	364
4 no fines concr. (incl. form)	0·792		4198	3068
5 l.w. aggr. concr. (incl. form)	0·288		1768	1309

Rigid sheets	k	12 mm	25 mm	50 mm
1 fibreboard	0·050	328		
2 expanded polystyrene	0·037	142	98	85
3 wood wool	0·082		298	236
4 straw board	0·086			239
5 cork board, 80 kg/m³	0·040	200	238	155
6 insulating plasterboard	0·060	433		
7 glass fibre slab, 16 kg/m³	0·037		59	46
8 glass fibre slab, 36 kg/m³	0·033	118	90	79

In situ finishes	k	12 mm	19 mm	25 mm
1 vermiculite plaster, walls	0·158	1210		
2 sprayed asbestos, soffits	0·040	423	363	323

In situ screeds	k	to falls, average: 50 mm	75 mm	100 mm
1 vermiculite, with topping	0·115	285	222	195
2 foam slag, with topping	0·197	441	387	331

Semi rigid mats and quilts	k	25 mm	38 mm	50 mm
1 glass fibre quilt	0·037	53	42	
2 glass fibre mat, paper faced	0·039	66	54	
3 glass fibre resin bonded	0·032	77	79	
4 mineral wool mat over joists	0·036	42	30	24
5 mineral wool resin bonded	0·036	135	115	104
6 mineral wool mat on concrete	0·033	62	50	44

Loose wool materials	k	50 mm	75 mm
1 slag wool	0·036	62	56
2 glass fibre	0·037	42	33
3 glass fibre	0·092	104	83
4 expanded mineral	0·043	52	43

Granular materials over ceiling			
5 expanded polystyrene	0·035	45	39
6 cork granules	0·036	50	46
7 vermiculite	0·062	46	41
8 perlite	0·048	35	31

Granular materials in cavity fill			
9 expanded polystyrene	0·035	53	
10 cork granules	0·036	58	
11 vermiculite	0·062	59	
12 perlite	0·048	42	

Foamed in situ cavity fill			
13 urea formaldehyde	0·030	54	42
14 polyurethane	0·035	91	102

* ie the cost divided by the resistance.

4.9
Ventilation

The term 'ventilation' is often used in a generic sense to include three different functions:

a the supply of fresh air
b the removal of heat from a space by air exchange

c physiological cooling

The last of these will be important in tropical situations, where convective cooling is insignificant, as the air is almost at skin temperature. The movement of air past the body surface will accelerate evaporation, thus give a significant cooling effect, even if the air temperature is slightly above skin temperature. At medium humidities an air velocity of 1 m/s could compensate for about 6 degC excess temperature. Here the requirement can be specified in terms of air velocity at the body surface, rather than in volumetric terms.

The first of the above functions is specified in volumetric terms and the requirement is relatively small. To remove CO_2 and other contaminants and to supply oxygen, the air exchange requirement can be specified in terms of m³/h.person, depending on the available volume per person.

Table 4.4   Ventilation requirements

if room volume m³/pers	rate of exchange m³/h. pers
0·28– 8·40	28
8·40–11·20	20
11·20–14·00	16
over 14·00	12

If the number of occupants is unknown, the requirement is set in number of air changes per hour. This varies between 1 and 3 for normal rooms, but goes up to 10 for commercial kitchens.

For a quick estimate the heat flow due to ventilation can be found by taking the volumetric specific heat of air as 0·36 Wh/m³ degC (1300 J/m³ degC), thus

$Q_v = 0.36 \times V \times N \times \Delta t$ where Q_v = vent. heat flow rate (W)
V = volume of space (m³)
N = number of air changes/h
Δt = temp. difference (degC)

(the product $V \times N$ is of course the ventilation rate in m³/h).

This expression is used to assess the heating requirement due to ventilation (the ventilation heat loss rate) in winter, but it can also be the basis of assessing the cooling effect of ventilation under overheated conditions. The latter is particularly relevant in moderate climates where solar radiation may cause overheating in buildings, when in fact the outdoor temperature does not exceed the comfort limit.

For example, if a 4 m² window admits solar radiation at the rate of 400 W/m², the solgar gain is

$Q_s = 400 \times 4 = 1600$ W

If the indoor temperature can be allowed to increase to 25°C when it is 18°C outdoors ($\Delta t = 7$°C) and we want to remove the solar gain by ventilation, thus Q_v should equal Q_s

$1600 = 0.36 \times V \times N \times 7$

$V \times N = \dfrac{1600}{0.36 \times 7} = 635$ m³/h

If the room is, say $5 \times 4 \times 2.4 = 50$ m³

we need $\dfrac{635}{50} = 12.7$ air changes per hour.

This is quite a high rate for mechanical ventilation, but can be readily achieved with adequate opening of windows. The foregoing would suggest that (perhaps apart from a few exceptional days) solar overheating would only occur in the UK if there is some reason preventing the opening of windows (eg a 'sealed' air conditioned building, or noise reasons).

A massive building would reduce the task of ventilation in removing the surplus heat, as it would store some of the heat gained during the day and dissipate it at night.

During a hot spell the night-time dissipation of heat from a massive building may be less than the day-time heat absorption. The temperature may show a variation such as that in Fig. 4.14, ie it may escalate due to an accumulation of heat in the fabric. [7] Under such conditions the simplest remedy would be to keep the room ventilated at night, thus to assist the heat dissipation, at least to an extent of keeping the daily mean temperature at the lower limit of comfort (say 16°C).

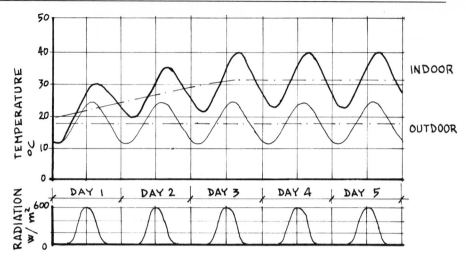

4.14
temperatures during a hot spell

Under winter conditions the heat lost from a building is the sum of the heat flowing through the enclosing elements (Q_c) and the ventilation heat loss rate (Q_v) (see above). With good insulation the Q_c may be reduced to such an extent, that Q_v becomes the dominant element.

In some cases (see section 5.19) the endeavour to reduce Q_v has lead to inadequate ventilation. If the minimum ventilation requirement is complied with, a significant amount of heat will be thrown away with the extracted (or expelled) warm used air, as this will be replaced by the fresh cold air. Ideally the used air should be thrown away but its heat content should be retained. Or, to put it in another way: the heat content of outgoing air should be transferred to the incoming air. Two solutions which do this, at least partially, are shown in Fig. 4.15. [8]

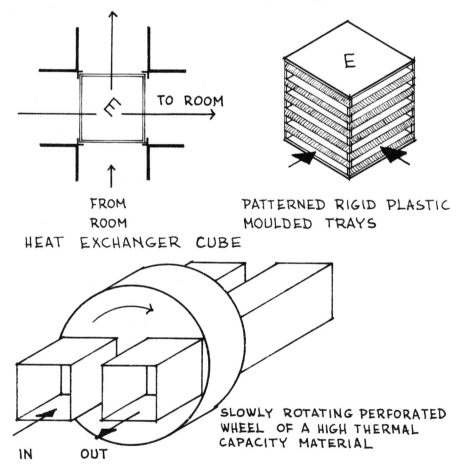

4.15
heat recovery in ventilation by static or
rotary air-to-air heat exchangers

References

1 Hardy, A C & O'Sullivan, P E
Insulation and fenestration
Oriel Press, 1967

2 Koenigsberger, O & Szokolay, S V
Climatic design
(Manual of tropical housing and building, part 1)
Longman, 1973

3 National Building Agency
The thermal and environmental benefits of improved thermal insulation (The EURISOL report)
NBA 1967/69

4 Petherbridge, P
Sunpath diagrams and overlays
HMSO, 1969

5 *The Building Regulations, 1972*
HMSO, 1972

6 Olgyay, V & Olgyay, A
Solar control and shading devices
Princeton Uni Press, 1957

7 Loudon, A G
Summertime temperatures in buildings
BRS Current Paper 47/1968

8 Pescod, D
Performance of air-to-air heat exchangers with moulded plastic plates
CSIRO (Australia)
Div Mech Eng Internal Report No. 121—1973

9 Page, J K
Solar energy and architecture
discourse given at the Royal Institution, 16 May 1974

Part 5   Solar Houses

5.1
The case for solar houses

Some 25% of all energy consumed is used for the heating of buildings and domestic hot water.

Space and water heating requires the lowest grade of energy.

The highest collection efficiencies are obtainable with low temperature collection.

From the juxtaposition of these three facts, the use of solar energy for space and water heating seems to be an obvious proposition.

Focusing devices require a tracking mechanism and respond to direct radiation only. Flat plate collectors can utilise both diffuse and direct radiation and may be fixed in one particular position. They may become part of the building envelope, replacing an enclosing element, such as a wall or a roof.

There are large areas involved. A domestic building may have a flat plate collector as much as 80 m² in area. It will not only dominate the appearance of the house, but will also impose quite strict design constraints. Any solution applying a solar heating system as an afterthought to a building already designed (let alone built) cannot hope to have the same degree of success as one which is an organic part of the design. Thus the discussion is focused on *solar houses,* rather than on *solar heating installations* in houses.

The following pages give a concise, factual description of a number of solar houses built in the last thirty years or so.

Some houses currently being built or for which the design work has been completed are also included.

Certain generalisations will be made and conclusions will be drawn at the end.

It should be noted that many of the systems and designs described in this chapter are protected by various patents. Anyone intending to use such a system or design should check the position regarding patent rights.

5.2
MIT solar house I

Built in 1939 as part of the Godfrey L Cabot solar energy conversion research project. Described by H C Hottel in *Proc. World Symposium on Applied Solar Energy,* Phoenix, 1955.

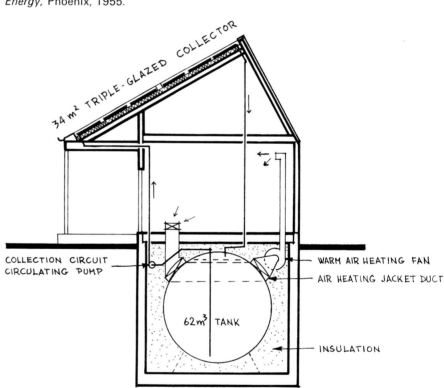

5.1
MIT solar house I: section

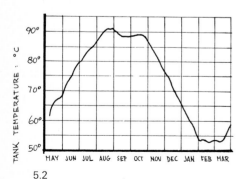

5.2
annual variation of temperature

* The term 'degree-days' is defined in section 7.2.

Building:	Single storey two-roomed laboratory (Fig. 5.1)	
Location:	Cambridge, Mass.	
	latitude:	42°N
	altitude:	approx 60 m
	degree-days* re 18°C:	3300
	design t_o:	−14°
	Jan av. radiation:	1860 Wh/m² day
Collector:	type:	water
	position:	south roof
	tilt:	28°
	area:	34 m²
	glazing:	triple
Storage:	type:	water
	volume:	62 m³
	heat capacity:	73 kWh/degC
	location:	in basement
	container:	steel tank

temperature: see Fig. 5.2
inter-seasonal storage attempted

Heating: collection circuit: water
distribution circuit: air
auxiliary: none
motors: pump and fan

Performance: see Fig. 5.2

Demolished in 1946 to make way for Solar House II.

5.3
MIT solar house III

EAST ELEVATION

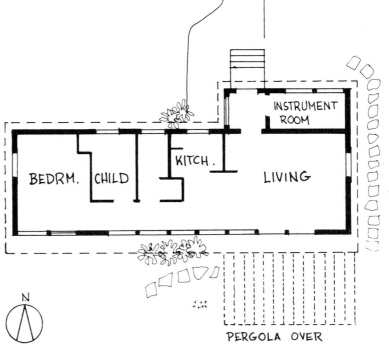

5.4
plan and elevation

5.3
MIT solar house II

Built in 1947 for testing purposes.

Building: single storey laboratory, consisting of one room and eight cubicles of 1·2 m width each
area: 55 m²

Collector: south vertical wall consists of black-faced tanks, some containing water, some with eutectic salts for latent heat storage.
glazing: partly double, partly triple

Later converted into Solar House III.

5.4
MIT solar house III

Built in 1948, using the shell of house II. (Figs. 5.3 and 5.4)

Designer: August L Hesselschwerdt

Building: single storey residence for a student couple with one child
floor area: 56 m²
spec. heat loss rate:* 185 W/degC
winter heat demand: 10 550 kWh

* The term 'specific heat loss rate' is defined in section 7.2.

Location: Lexington, Mass.
  latitude: 42°N
  altitude: approx 60 m
  degree-days re 18°C: 3300
  Jan av. radiation: 1860 Wh/m² day

Collector: type: water
  position: south roof
  tilt: 57°
  area: 37 m²
  construction: 0·5 mm plate, 10 mm tubes at 150 mm crs, 20 mm headers all copper. Connected to manifold by rubber hoses (in 15 panels)
  glazing: double

Storage: type: water
  volume: 4·5 m³
  location: in roof space
  container: steel tank
  relative capacity: 120 litre/m²

Heating: collection circuit: water to storage
  distribution circuit: water from storage
  emitters: 20 mm pipes at 230 mm crs embedded in plaster ceiling (Fig. 5.5)

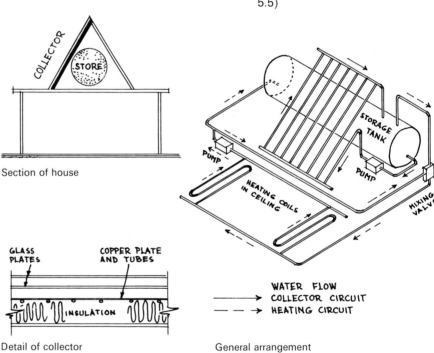

Section of house

Detail of collector

General arrangement

GLASS PLATES

COPPER PLATE AND TUBES

INSULATION

WATER FLOW
COLLECTOR CIRCUIT
HEATING CIRCUIT

5.5
details of the system

  motors: 373 W in collection
                               125 W in distribution
  auxiliary: electric
  — capacity: 4 kW
  — connection: independent

Hot water: none

Performance: period: winter 1951/52 (Oct–Apr)
  av. collect efficiency: 43%
  do. re total radiation: 31%
  auxiliary heat used: 1899 kWh
  solar contribution: 82% (49% by system, 33% by windows)

Notes: south wall 47% window (17 m²), triple glazed with accordion type shutters

  collection temperature in winter 6 months: min 38°C, max 71°C. Minimum circulation temperature for heating set at 29°C

Collector drained to prevent freezing and to reduce heat capacity loss.

Reflective flat roof on south did not prove to be beneficial.

The house burnt down due to the malfunctioning of a paraffin heater.

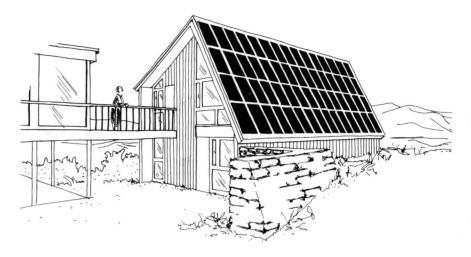

5.6
MIT solar house IV

5.5
MIT solar house IV

Designed in 1954–55, scheduled for completion in 1956, completed in 1959, as part of the Godfrey L Cabot solar energy conversion research project.

Designer:	MIT Dept of Mechanical Engineering C D Engebretson	
Building:	two-storey residence (Fig. 5.6)	
	floor area:	134 m²
	spec. heat loss rate:	263 W/degC
	max. heat load:	10·8 kW
	winter heat demand:	20 135 kWh (Oct–April)
Location:	Lexington, Mass.	
	latitude:	42°N
	altitude:	approx 60 m
	degree-days re 18°C:	3300
	design t_o:	−14°C (−23°C for max heat load)
	Jan av. radiation:	1860 Wh/m² day
Collector:	type:	water
	position:	south roof/wall
	tilt:	60°
	area:	60 m²
	construction:	alum. sheets, clip-in copper tubes (Fig. 5.8)
	glazing:	double
Storage:	type:	water
	volume:	5·67 m³
	container:	steel tank
	heat capacity:	117 kWh (over 18 degC) = 6·5 kWh/degC
	temperature:	49°C normal max
Heating:	collection circuit:	water, 2430 lit/h
	distribution circuit:	water, 817 lit/h
	emitters:	fan-convector 0·4 m³/s
	motors:	2 pumps + 1 fan
	auxiliary:	oil fired boiler
	— capacity:	12 kW
	— connection:	exchanger in storage

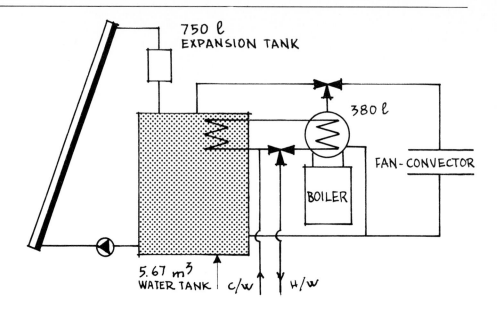

5.7
system diagram

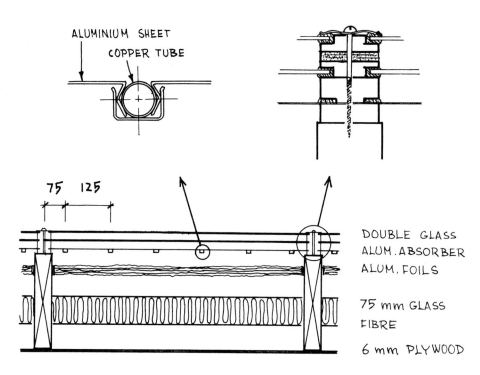

5.8
collector details

Hot water:	pre-heating, through coil in storage (Fig. 5.7) topping-up through coil in auxiliary tank	
Performance:	period:	winter 1960/61 (Oct–Mar)
	av. collect efficiency:	40·8%
	do. re total radiation:	32·6%
	auxiliary heat used:	8468 kWh (10 665 kWh inc. hot water)
	solar contribution:	11 163 kWh (13 888 kWh inc. hot water) 56·8%
Notes:	suggested that 1 m² collector equals in annual output 40 litre fuel oil	
	economic lower limit of storage temperature for heating: 28°C	
	Because of the onerous maintenance problems (glass breakages, leaks and the corrosion of aluminium) the solar heating system was abandoned after two years of use.	

5.6
Dover house

Built in 1948	
Designers:	Telkes, Raymond and Peabody
Building:	two-storey, three-bedroom residence (Figs. 5.9 and 5.10)
	floor area: 135 m²

5.9
Dover house

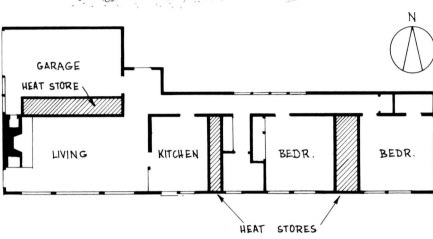

5.10
plan

	floor:	vermiculite concrete on 300 mm gravel fill, with 50 mm cork edge insulation
	walls:	stud frame, 100 mm mineral wool insulation
	ceiling:	100 mm mineral wool insulation on 'Insulite' panels
	roof:	corrugated aluminium
Location	Dover, Massachusetts	
	latitude:	42°N
	altitude:	approx 60 m
	degree-days re 18°C:	3300
	design t_o:	−14°C
	Jan av. radiation:	1860 Wh/m² day
Collector:	type:	air
	position:	south wall
	tilt:	90° (vertical)
	construction:	sheet metal panels
	glazing:	double
Storage:	type:	latent heat of fusion: Glauber salt (sodium sulphate) (Fig. 5.11)
	volume:	13 m³ in three sections
	container:	metal canisters
	heat capacity:	9 days heat requirement
Heating:	collection circuit:	air to storage
	distribution circuit:	air from storage
	auxiliary:	none
Notes:	stratification of storage compounds has progressively increased	
	heat transfer from storage to air was inadequate	
	after some five years of use, the solar heating system was removed because of its gradually deteriorating performance	

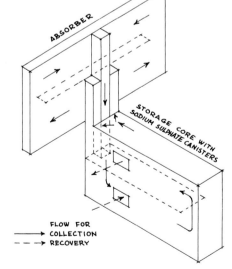

5.11
system isometric (one of three
independent units)

5.7
Boulder house

Conversion of an existing five-roomed bungalow, completed in 1945. Project supported by the American Window Glass Co of Pittsburgh, Pennsylvania.

Designer: George O G Löf

Building: single storey bungalow (Fig. 5.12)

5.12
Boulder house

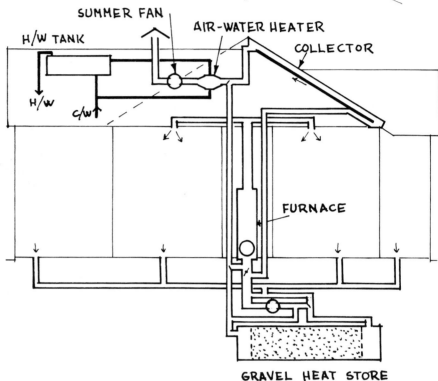

5.13
diagrammatic section

	floor area:	93 m²
	winter heat demand:	24 220 kWh
Location:	Boulder, Colorado	
	latitude:	40°N
	degree-days re 18°C:	3056
Collector:	type:	air
	position:	south roof
	tilt:	27°
	area:	43 m²
	construction:	overlapped glass slats
Storage:	type:	20 mm gravel
	volume:	5 m³ (8 tonne)
	location:	in basement
	container:	cinder block bin
Heating:	collection circuit:	air to gravel or to room direct
	distribution circuit:	house air through gravel
	auxiliary:	gas furnace
	— connection:	heats air from collector or from storage (Fig. 5.13)
Hot water:	pre-heating, summer only, through finned tube heat exchanger gravity water circulation, 300 litre tank	
Performance:	period:	Sept–May
	auxiliary heat used:	18 000 kWh (gas)
	solar contribution:	6220 kWh = 26%
Notes:	with smooth operation the contribution would be 55% (due to glass breakages and to repair and modification of the control equipment, the system was out of service for more than half the time of the period).	

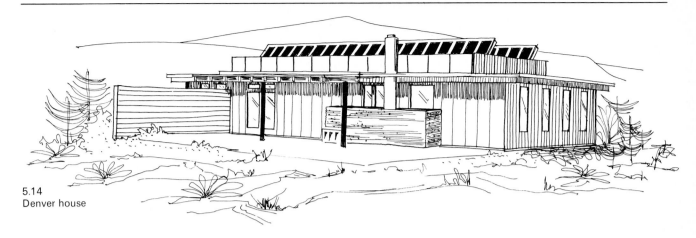

5.14
Denver house

5.8
Denver house

Built in 1956–57 as part of the American St Gobain Glass Corporation's solar energy development programme.

Designers:	George L G Löf, M M el Wakil, J P Chiou	
Building:	single storey residence (Fig. 5.14)	
	floor area:	297 m²
	spec. heat loss rate:	704 W/degC
	max. heat load:	31 kW
	winter heat demand:	56 900 kWh
Location:	Denver, Colorado	
	latitude:	40°N
	altitude:	1646 m
	degree-days re 18°C:	3390
	design t_o	-18°C
Collector:	type:	air (Fig. 5.16)
	position:	mounted on flat roof
	tilt:	45°
	area:	56 m²
	construction:	overlapped glass slats
	glazing:	half single, half double

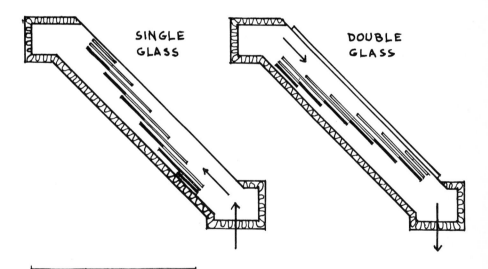

AIR FLOW:
⟶ FOR COLLECTION
-- → FOR RECOVERY

5.15
system principles (see also Fig. 3.18)

5.16
collector detail

TYPICAL GLASS LAMELLA

Storage:	type:	40 mm crushed rock
	volume:	7 m³ (11 tonne)
	location:	in centre of house
	container:	two fibre cylinders
	heat capacity:	88 kWh
	temperature:	60°C normal max
Heating:	collection circuit:	air, 1300 m³/h
	distribution circuit:	air, 1300 m³/h
	motors:	930 W motor to fan

	auxiliary:	natural gas furnace
	— capacity:	30 kW
	— connection:	heats air from collector or from storage (Fig. 5.15)

Hot water: pre-heating through air-to-water heat exchanger topping up by automatic gas heater

Performance: period: winter 1959/60

	av. collect efficiency:	34·6%
	do. re total radiation:	24·5%
	auxiliary heat used:	41 590 kWh (6500 m³ gas)
	solar contribution:	15 160 kWh = 26·5%

5.9
Princeton house

Built in 1956 by the Curtiss Wright Corporation for the purposes of testing various solar devices.

Designer: Aladar Olgyay

Building: a single storey laboratory, with a terrace between two blocks (Fig. 5.17)

	floor area:	110 m²
	spec. heat loss rate:	198 W/degC
	max. heat load:	7·33 kW

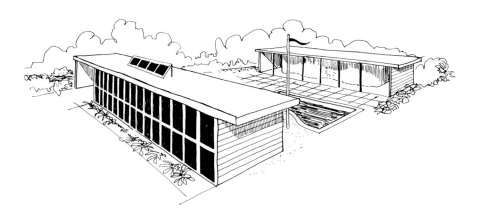

5.17
Princeton house

Location:		
	Princeton, New Jersey	
	latitude:	40°N
	altitude:	15 m
	degree-days re 18°C:	2830
	design t_o:	−12°C

Collector:		
	type:	air
	position:	south wall
	tilt:	90° (vertical)
	area:	56 m²
	construction:	metal sheet
	glazing:	double

Storage:		
	type:	latent heat of fusion
	volume:	7·8 m³
	container:	canisters in bin
	heat capacity:	733 kW

Heating:		
	collection circuit:	air to storage
	distribution circuit:	air from storage
	auxiliary:	none

Hot water: independent solar water heating system

Performance:		
	period:	December–January
	av. collect efficiency:	46·5%

Notes: if extra cost of installation is not more than 2·3 times the cost of a conventional heating system, the break-even point will be less than 20 years (discounted at 4·5% interest)

recommends: collector = 0·5 × floor area
storage: 5–6 days requirement

5.10
Albuquerque office

Built in 1956

Designers:	F H Bridges, D D Paxton and R W Haines
Building:	single storey office block (Fig. 5.18)
	floor area: 410 m²
	spec. heat loss rate: 893 W/degC
	winter heat demand: 47 730 kWh (Oct–May)

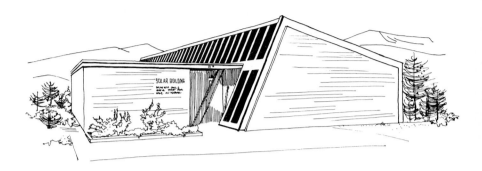

5.18
Albuquerque office

Location:		Albuquerque, New Mexico
	latitude:	35°N
	degree-days re 18°C:	2230
Collector:	type:	water
	position:	south wall
	tilt:	60°
	area:	71 m²
	construction:	13 mm copper tubes soldered to back of aluminium plates
	glazing:	single
Storage:	type:	water
	volume:	23 m³
	container:	steel tank
	location:	underground
Heating:	collection circuit:	water to storage
	distribution circuit:	water from storage or through auxiliary tank
	emitters:	ceiling panels in main block, floor coils in flat-roofed block + fan-coil in ventilators
	auxiliary:	heat pump
	— capacity:	26 kW (heat delivery)
	— connection:	through auxiliary tanks (Fig. 5.19)

Cooling:	**1**	evaporative cooling, short circuited whilst heat is collected in tank for night use (spring/autumn operation)
	2	evaporative cooling of space and storage (summer, day and night operation)
	3	heat pump cooling (in heat waves) (Fig. 5.19)

Performance:	period:	winter 1956/57
	solar contribution:	
	— direct:	29 933 kWh = 62·7%
	— through heat pump:	17 797 kWh = 37·3%
	electricity consumed:	3920 kWh = 8·2%
	coeff. of performance:	4·5

Notes:	originally aluminium tube-in-strip collectors were used, but developed pinhole corrosion. Copper tubes were soldered to the back of the original aluminium panels.
	Collector is drained under freezing conditions.

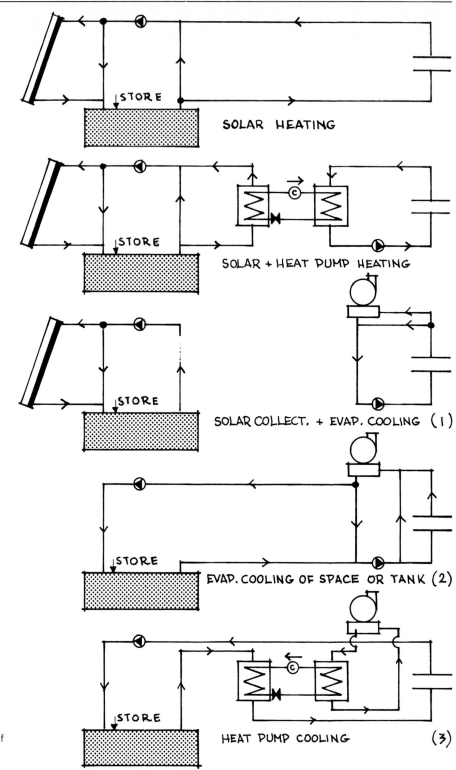

SOLAR HEATING

SOLAR + HEAT PUMP HEATING

SOLAR COLLECT. + EVAP. COOLING (1)

EVAP. COOLING OF SPACE OR TANK (2)

HEAT PUMP COOLING (3)

5.19
system diagram showing the five modes of
operation (non-working parts in each
mode omitted for clarity)

5.11
Rickmansworth house

Completed in 1956 for the designer's own use

Designer: Edward J W Curtis

Building: two-storey residence (Fig. 5.20)
               floor area: 136 m²
               spec. heat loss rate: 415 W/degC

Location: Rickmansworth (nr London)

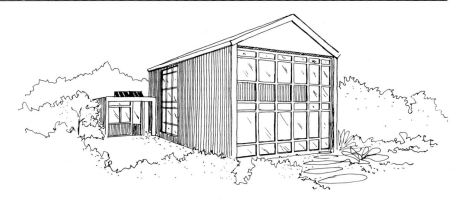

5.20
Rickmansworth house

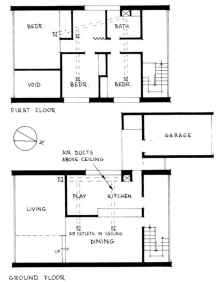

FIRST FLOOR

GROUND FLOOR

5.21
plans

5.12
Nagoya laboratory

latitude:		52°N
degree-days re 18°C:		2600
Jan av. radiation:		590 W/m² (horizontal)
		977 W/m² (vertical S)

Collector: window wall facing south
area: approx 28 m²
glazing: 'Plyglass' sealed double glazing units

Heating: a 'passive' system, relying on solar heat gain through windows, which are covered by heavy curtains during no-gain periods

auxiliary: heat pump
— capacity: 12 kW (heat delivery)

first used with a fan coil heat exchanger mounted on the annexe roof, acting as the heat source. Later modified to a water source.

Cooling: the same system is used for air cooling in summer

Performance: no measured data is available, except the designer's (user's) statement that it has performed satisfactorily for 18 years, with an average annual electricity consumption of:

day tariff: 8590 kWh
night tariff: 1509 kWh

A solar energy research building, for the Government Industrial Research Institute, Agency for Industrial Science and Technology of Japan, completed in 1958.

Designer: Nobuhei Fukuo et al.

Building: a two-storey laboratory block (Fig. 5.22) with one level heated
floor area: (heated) 82 m²
spec. heat loss rate: 814 W/m²

5.22
Nagoya laboratory

Location: Nagoya, Japan
latitude: 35°N
altitude: 60 m
Jan. av. radiation: 3200 Wh/m² day

Collector: type: water
position: on flat roof
tilt: 35°
area: 28 m²
construction: aluminium tube-in-strip
glazing: none

Storage:	type:	water
	volume:	$2 \times 5 \cdot 6\ \text{m}^3$
	container:	underground tanks
	heat capacity:	64 kWh each
	temperature:	13°C max normal
Heating:	collection circuit:	warmed water to store
	— flow rate:	3600 litre/h
	distribution circuit:	hot water to air heater
	motors:	2 pumps
	auxiliary:	heat pump
	— capacity:	2·2 kW compressor motor
	— connection:	heat transfer from solar storage tank to hot tank
Hot water:	none	
Cooling:	disposal:	night radiation and convection
	auxiliary:	heat pump (same as above)
	space cooling:	cool water to air heat-exchanger used for heating
Performance:	period:	1958/59 (Nov–March)
	av. collect efficiency:	54% (in January)
	auxiliary fuel:	electricity
	— amount used:	1370 kWh = 25%
	solar contribution:	4245 kWh = 75%

5.13
Tucson laboratory

A solar energy laboratory building for the Institute of Atmospheric Physics of the University of Arizona, completed in 1959. A sealed, air conditioned building where cooling is effected by chilled water circulated in ceiling panels.

Designer:	Raymond W Bliss
Building:	a single-storey laboratory building (Figs. 5.23 and 5.24)
	floor area:	134 m²
	spec. heat loss rate:	396 W/degC
	max. heat load:	10·26 kW
	winter heat demand:	10 353 kWh

5.23
Tucson laboratory

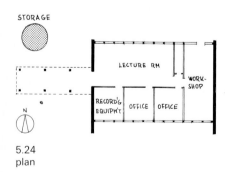

5.24
plan

Location:	Tucson, Arizona	
	latitude:	32°N
	altitude:	792 m
	degree-days re 18°C	1000
	Jan av. radiation:	approx 3490 Wh/m² day
	design t_0:	−1°C
Collector:	type:	water
	position:	on south roof
	tilt:	7°
	area:	150 m²
	construction:	expanded copper tubed sheet (Fig. 5.26)
	glazing:	none
Storage:	type:	water
	volume:	17 m³
	location:	outside building
	container:	tank in two sections

	heat capacity:	approx 300 kWh
	temperature:	43°C normal max
Heating:	collection circuit:	water to storage
	distribution circuit:	water from storage
	emitters:	radiant ceiling 123 m²
	motors:	two pumps, 185 W each
	auxiliary:	heat pump
	— capacity:	370 W compressor motor
	— connection:	transfer heat from one tank to the other (Fig. 5.25)

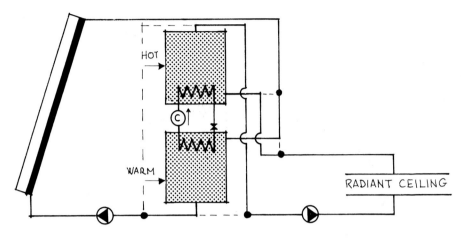

HEATING MODE

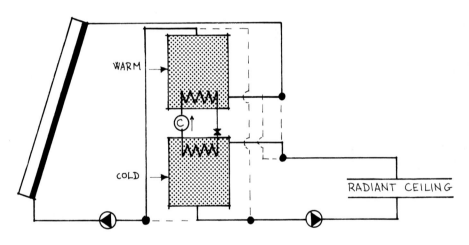

COOLING MODE

5.25
system diagram showing the two modes
of operation

Hot water:	none	
Cooling:	disposal method:	night radiation and convection
	auxiliary:	heat pump (same as used for heating)
	space cooling:	chilled water in ceiling panels
Performance:	period:	winter 1959/60
	av. collect efficiency:	28·7% (in Jan only)
	do. re total radiation:	17·2% (in January only)
	auxiliary heat used:	1470 kWh, electricity
	solar contribution:	8887 kWh = 86%

5.26
detail of roof collector/sink

5.14
Tokyo house

A private residence, built by the designer for his own use, completed in 1958, after his first experimental house built in 1956 was destroyed by fire

Designer:	Masanosuke Yangimachi	
Building:	a two-storey residence (Figs. 5.27 and 5.28)	
	floor area:	228 m²
	spec. heat loss rate:	374 W/degC
	max. heat load:	7·33 kW
	winter heat demand:	18 770 kWh

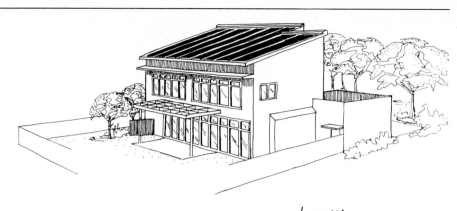

5.27
Tokyo house

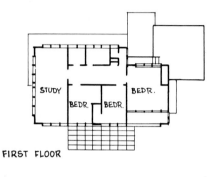

FIRST FLOOR

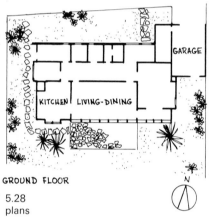

GROUND FLOOR

5.28
plans

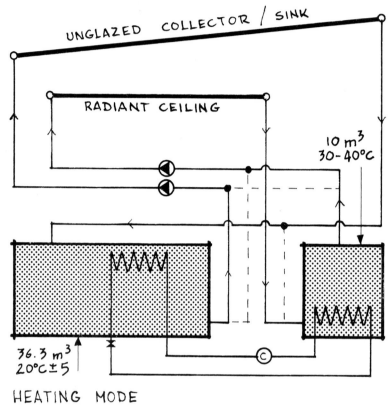

UNGLAZED COLLECTOR / SINK

RADIANT CEILING

10 m³
30-40°C

36.3 m³
20°C±5

HEATING MODE

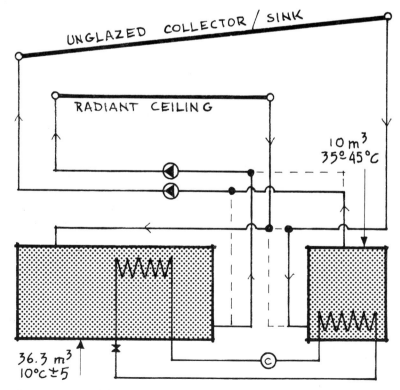

UNGLAZED COLLECTOR / SINK

RADIANT CEILING

10 m³
35°-45°C

36.3 m³
10°C±5

COOLING MODE

5.29
system diagram showing the two modes
of operation

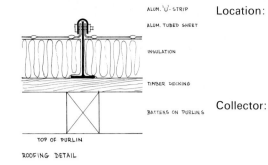

ALUM. 'U'- STRIP
ALUM. TUBED SHEET
INSULATION
TIMBER DECKING
BATTENS ON PURLINS

TOP OF PURLIN
ROOFING DETAIL

600 mm WIDE UNITS CONNECTED TO
50 mm ∅ HEADER PIPE BY HOSES

ALUMINIUM TUBED SHEET

5.30
details of roof collector/sink

Location:		
	Tokyo, Japan	
	latitude:	36°N
	altitude:	60 m
	degree-days re 18°C:	2110
	Jan av. radiation:	2650 Wh/m² day
	design t_0:	0°C
Collector:	type:	water
	position:	south roof
	tilt:	15°
	area:	130 m²
	construction:	expanded aluminium tubed sheet (Fig. 5.30)
	glazing:	none
Storage:	type:	water
	volume:	36·3 m³ + 10 m³
	container:	concrete tanks
	location:	in basement
	heat capacity:	704 kWh (main tank)
	temperature:	39°–44°C normal max
Heating:	collection circuit:	warmed water to store
	distribution circuit:	hot water from small tank (Fig. 5.29)
	emitters:	radiant ceiling + fan convector
	motors:	3 pumps, 370 W each + 1 pump 745 W
	auxiliary:	heat pump
	— capacity:	2·2 kW compressor motor
	— connection:	transfer heat from main tank to small tank

Hot water: separate system with 32 m² collector and a heat pump with a 745 W compressor motor

Cooling:	disposal method:	night radiation and convection
	auxiliary:	heat pump (same as used for heating)
	space cooling:	the same radiant ceiling + fan convector
Performance:	period:	winters 1959–60–61
	av. collect efficiency:	22%
	auxiliary heat used:	5580 kWh electricity
	solar contribution:	13 200 kWh = 70%

5.15
Thomason house I

Private residence, built by the designer for his own use in 1959

Designer:	Harry E Thomason	
Building:	single storey, three-bedroom residence (Fig. 5.31)	
	floor area:	100 m²
	spec. heat loss rate:	154 W/degC
	Jan av. heat load:	2·5 kW
Location:	Washington, DC	
	latitude:	40°N
	altitude:	15 m
	degree-days re 18°C	2390
	Jan. av. radiation:	1860 Wh/m² day
	design t_0:	−10°C
Collector:	type:	water, free flow down ('trickle'-type)
	position:	south wall and roof
	tilt:	60° and 45°
	area:	78 m²
	construction:	corrugated aluminium
	glazing:	single (the 0·13 mm Mylar plastic sheet second cover failed after 2 years)
Storage:	type:	water and rock

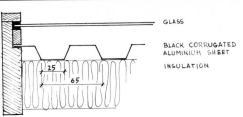

BARGE DETAIL

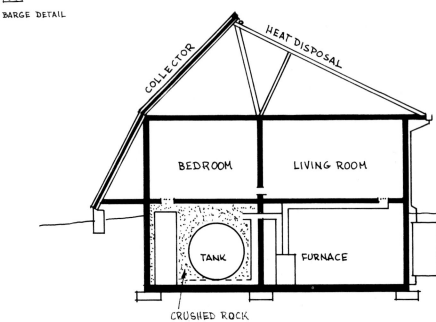

5 31
Thomason house I, diagrammatic section
and roof edge detail

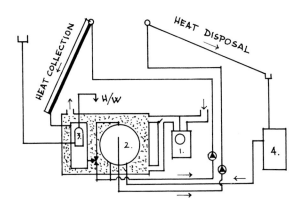

5.32
system diagram

	volume:	6 m³ water + 45 tonne rock
	container:	steel tank + conc. bin
	location:	in basement
	heat capacity:	470 kWh
	temperature:	52°–57°C normal max
Heating:	collection circuit:	water to store + cond. to rock
	— flow rate:	1620 litre/h
	distribution circuit:	air through rock
	motors:	1 pump + 1 fan
	auxiliary:	oil furnace
	— connection:	warming air, bypassing storage (Fig. 5.32)
Cooling:	disposal method:	night radiation, convection, evaporation
	auxiliary:	none
	motors:	186 W pump motor
	space cooling:	air circulated through cooled rocks
Performance:	period:	winter 1960/61
	auxiliary fuel used:	160 litre oil
	— heat produced:	1320 kWh
	solar contribution:	95% (estimated)

5.16
Thomason house II

Private residence, built by the designer for the use of his children, completed in 1960

Designer: Harry E Thomason

Building: single-storey two-bedroom residence (Fig. 5.33) with outdoor swimming pool

floor area: 64 m²

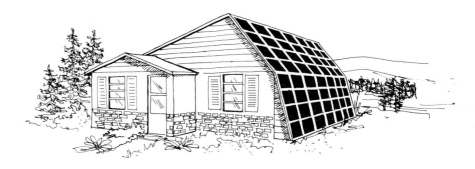

5.33
Thomason house II

Location: as above (built on the same site)

Collector: type: water, trickle type
                 position: south wall and roof
                 tilt: 65° and 55°
                 area: 53 m²

Storage: as above, but 54 tonne rock

Heating: as above, but the auxiliary heater is an electric resistance element

5.17
Thomason house III

Private residence, built by the designer for his own use, completed in 1963

referred to as the 'Solaris' system

Building: single-storey, four-bedroom residence (Fig. 5.34)

floor area: 142 m²+swimming pool, game room, etc.

5.34
Thomason house III

Location: as above

Collector: type: water, trickle type
                 area: 85 m²
                 position: south roof
                 tilt: 60°

Storage: as above, but 6 m³ water+45 tonne rock and indoor swimming pool acts as secondary storage

Note: flat roof over swimming pool is reflective to assist the collector (Fig. 5.35)

5.18
Capri laboratory

A building for the Swedish astrophysical station in Capri, Swedish Academy of Science, built in 1960

Designers: G Pleijel and B Lindstrom

Building: two-storey laboratory building (Fig. 5.36)
                 floor area: 180 m²
                 spec. heat loss rate: 264 W/degC
                 winter heat demand: 5256 kWh

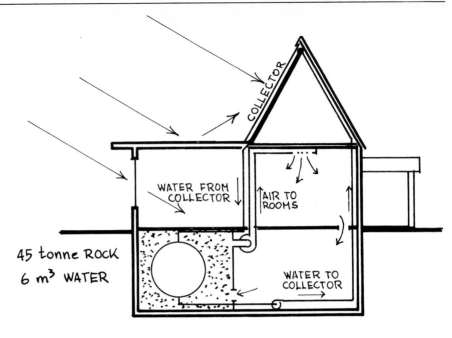

45 tonne ROCK
6 m³ WATER

5.35
operating principles

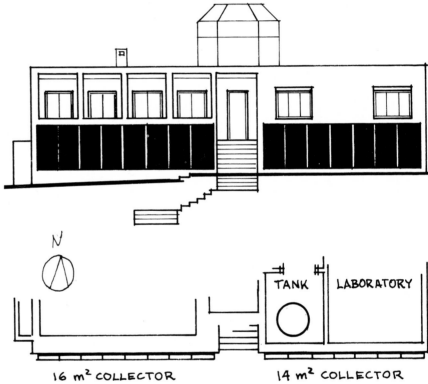

5.36
Capri laboratory, south-west elevation and
part basement plan

16 m² COLLECTOR   14 m² COLLECTOR

Location:	Capri, Italy	
	latitude:	41°N
	altitude:	228 m
	degree-days re 18°C:	1500 (approx)
	design t_o:	−1°C
Collector:	type:	water
	position:	south-west wall
	tilt:	90° (vertical)
	area:	30 m²
	construction:	steel radiator panels
	glazing:	one sheet glass (outside) one sheet Teslar plastic (Fig. 5.38)
Storage:	type:	water
	volume:	3 m³
	container:	steel tank
	location:	in basement
	temperature:	35°–58°C
Heating:	collection circuit:	heated water to storage

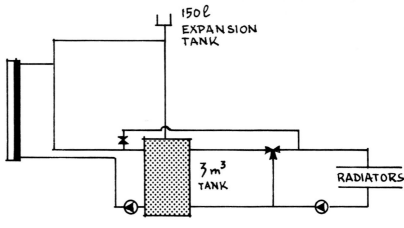

5.37
system diagram

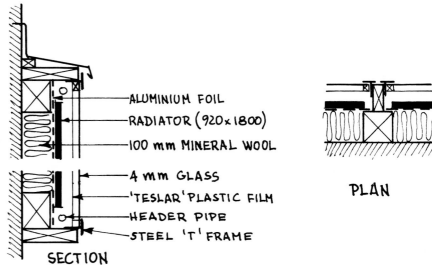

— ALUMINIUM FOIL
— RADIATOR (920×1800)
— 100 mm MINERAL WOOL

— 4 mm GLASS
— 'TESLAR' PLASTIC FILM
— HEADER PIPE
— STEEL 'T' FRAME

SECTION

PLAN

5.38
collector details: section and plan

— flow rate:	480 litre/h
distribution circuit:	heated water from storage
— flow rate:	480 litre/h (Fig. 5.37)
emitters:	panel radiators
motors:	two pumps
auxiliary:	stove + electric resistance element
— connection:	independent, direct room heating

Hot water: independent system, with separate collector

Performance: predicted only
solar contribution: 70%

5.19
Wallasey school

A new wing, added to the existing (1954) building of St George's Secondary School, completed in 1961

Designer: A E Morgan

Building: two blocks, both two-storey (Figs. 5.39 and 5.40)
floor area—east block: 497 m²
—west block: 870 m²
(here the west block is described in detail)

Location: Wallasey, near Liverpool
latitude: 53°N
mean annual t_o: 10°C
degree-days re 18°C: 2500

Collector: the whole of the south wall (window)
area: 500 m²
construction:
— outer skin: all glass
— inner skin: 12% of area black painted masonry 88% glass
33% of area has reversible aluminium panels: bright side out Apr–Oct, black side outwards Nov–March.

5.39
Wallasey school

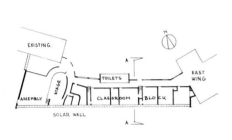

5.40
plan and section

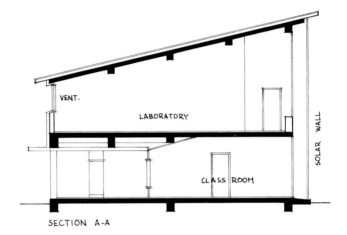

SECTION A-A

Storage: the building fabric:
180 mm concrete roof and 230 mm brick walls, with 125 mm polystyrene insulation on the outside of the mass

Heating: solar radiation, lighting and occupants, thermal control: by ventilation only

Performance: for over 20 years, without any auxiliary heating the following temperatures have been maintained:
June–Oct: 18°–24°C
Nov–Feb: 16°–20°C
Mar–May: 17°–21°C
MRT (mean radiant temp) normally 0·5°C above DBT (dry bulb temp)

5.20
Brisbane houses

An experimental project, designed in 1959 by the Department of Architecture and Department of Mechanical Engineering of the University of Queensland

Designers: Norman R Sheridan (mechanical)
M Juppenlatz (architect)

5.41
Brisbane house design

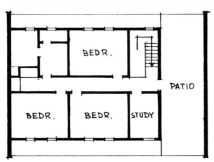

FIRST FLOOR

AIR CONDITIONING EQUIPMENT

GROUND FLOOR

N

5.41/a
plans

* Dry bulb temperature.
† Wet bulb temperature.

Building:	two-storey residence (Figs. 5.41 and 5.41/a)	
	floor area:	2×66 m²
	cooling load:	7 kW max
		73 kWh/day
	design t_i:	27°C DBT*
		23°C WBT†

Location:	Brisbane, Australia	
	latitude:	27°S
	design t_o:	38°C DBT
		33°C WBT
	summer av. radiation:	5020 Wh/m² for the daily 6 h operating period

Collector:	type:	water
	position:	north roof
	tilt:	5°
	area:	76 m²
	construction:	copper plate and tubes
	glazing:	double
	collect. temperature:	82°C

Cooling:	absorption refrigeration	
	refrigerant:	water
	absorbent:	lithium bromide
	generator temperature:	71°–82°C
	evaporator temperature:	+4·5°C

generator (separator) is operated by hot water from storage, thus after sunset cooling is possible

	motors:	two pumps+fan total 400 W
	distribution:	ducted cooled air (recirculated)

Performance:	(estimated)	
	av. collect efficiency:	35%
	coefficient of performance of absorption refrigerator:	0·55

Note: this project has not been built. Subsequent re-thinking of the design lead to the construction of the following house which was completed in 1966.

Designers: Norman R Sheridan (mechanical)
W H Carr (architect)

Building:	single-storey, three-bedroom residence (Fig. 5.42)	
	floor area:	132 m²
	base and roof area:	300 m²
	construction:	concrete base slab, brick end walls, timber portals support roof of 0·36 mm copper on plywood. Inside box (walls and roof) of 75 mm polystyrene between two sheets of plywood

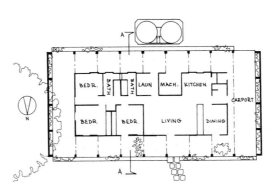

SECTION A-A

5.42
Brisbane house:
isometric, system details, plan and section

Location: Moggill (Brisbane) Australia
latitude: 27°S
climate: as above

Collector: type: water
position: north roof
tilt: 10°
area: 69 m² (12 panels of $1·2 \times 4·8$ m²*)
construction: 1 mm copper sheet
with 13 mm copper tubes at 150 mm crs on 100 mm polyurethane insulation, selective black surface
glazing: double with 25 mm spacing (in $1·2 \times 1·2$ m panes)

Storage: hot, in collection circuit, water
— volume: 320 litre per panel
cold, in distribution circuit, rock pile (20 mm river gravel)
—mass: 30 tonnes
— container: two steel tanks

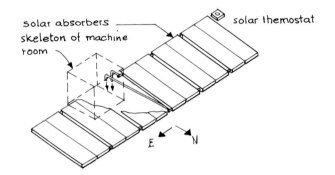

* In fact only one of the twelve panels has been installed, with an electrically operated amplifier simulating the performance of the remaining eleven panels.

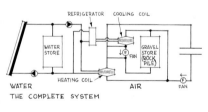

THE COMPLETE SYSTEM

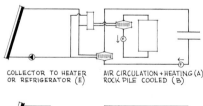

COLLECTOR TO HEATER OR REFRIGERATOR (E)   AIR CIRCULATION + HEATING (A) ROCK PILE COOLED (B)

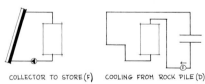

COLLECTOR TO STORE (F)   COOLING FROM ROCK PILE (D)

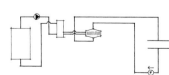

STORE TO REFRIGERATOR (G)   COOLING BY REFRIGERATOR (C)

5.42/a
system diagram, showing the various modes of operation

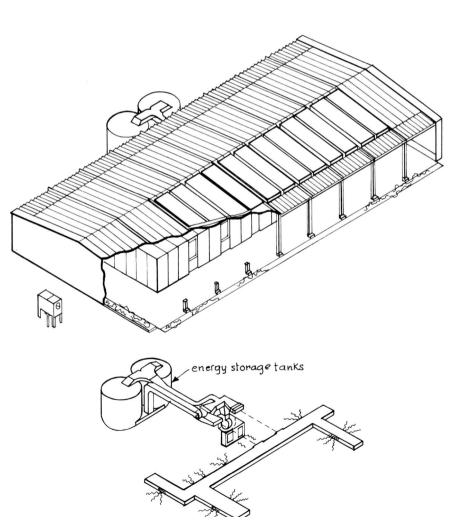

Cooling: absorption refrigerator, 10·5 kW rating (designed to operate with 100°C steam, used here with 60–96°C water)

refrigerant:	water
absorbent:	lithium bromide
condenser:	water cooled
cooling tower:	cross-flow type

Distribution: recirculated room air cooled by direct contact evaporator coil (Fig. 5.42/a)

Heating: separate heating coil fed from water storage or direct from collector

Hot water: separate solar hot water system

Performance:
collection efficiency:	50% instantaneous
	30% over operating day
generator starting temperature:	90°C
re-start:	84°C
operating minimum:	60°C

using hot water of 93°C it gives a coeff. of performance of 0·7 and a cooling rate of 8·8 kW

the system can maintain an internal temperature of 25°C when the outdoor temperature is 38°C

5.21
Toronto house

Designed in 1960 at the University of Toronto, Department of Mechanical Engineering, to study the possibility of inter-seasonal storage of heat (by using analogue simulation techniques)

Designers: E A Allcutt and F C Hooper

Building: two-storey residence (Fig. 5.43)
floor area:	114 m²
spec. heat loss rate:	408 W/degC

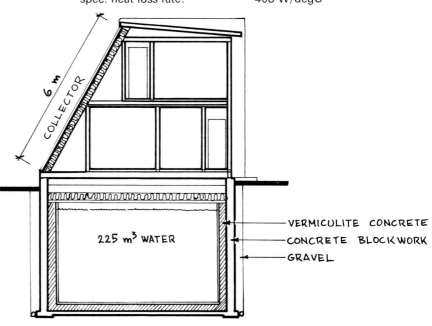

5.43
Toronto house: section

Location: Toronto, Canada
latitude:	43°N

Collector:
type:	water
area:	57 m²
position:	south wall
tilt:	60°

Storage:
type:	water
volume:	225 m³
location:	in basement
container:	concrete and blockwork
temperature:	63°C (max reached)
	27°C (min useable)
heat capacity:	261 kWh/degC
for above range:	9400 kWh

Performance: results of a three-year simulation are given in Fig. 5.44

Note: storage tank is not insulated, thus surrounding ground becomes an extension to the store

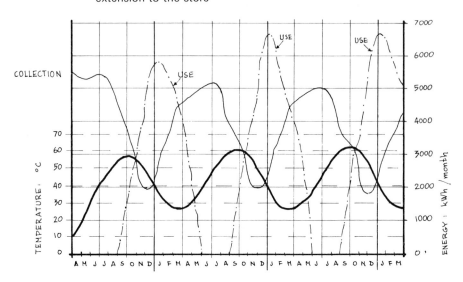

5.44
performance prediction

5.22
Chauvency-le-Château house

Completed in 1972

Designers: J Michel and A F Trombe with Felix Trombe, consultant

Building: single storey, five-roomed residence (Figs. 5.45 and 5.46)
floor area: 106 m²

5.45
Chauvency-le-Château house

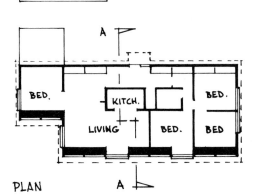

PLAN

STORAGE MASS,
BLACK SURFACE,
BEHIND DOUBLE GLASS

SECTION A-A

5.46
plan and section

Location:	Chauvency-le-Château, Meuse, France
	latitude:	50°N

Collector:	type:	air+mass
	area:	45 m²
	position:	south wall
	tilt:	90° (vertical)
	glazing:	double

Heating: radiation heats black surface of massive concrete wall. Part of this heat is stored, part is distributed to rooms by gravity circulation of air. Opening outside vents, the hot air is discharged, providing an increased ventilation. (See also Fig. 3.6.)

auxiliary: electric resistance heaters
— connection: independent

5.23
Odeillo houses

A group of houses designed in 1973, to be erected at Odeillo, near the Centre National de la Recherche Scientifique solar laboratories and solar furnace

Designer: Jacques Michel

Buildings: 31 residences in two- and four-storey blocks (Fig. 5.47)
floor area: 100 m² av. per unit

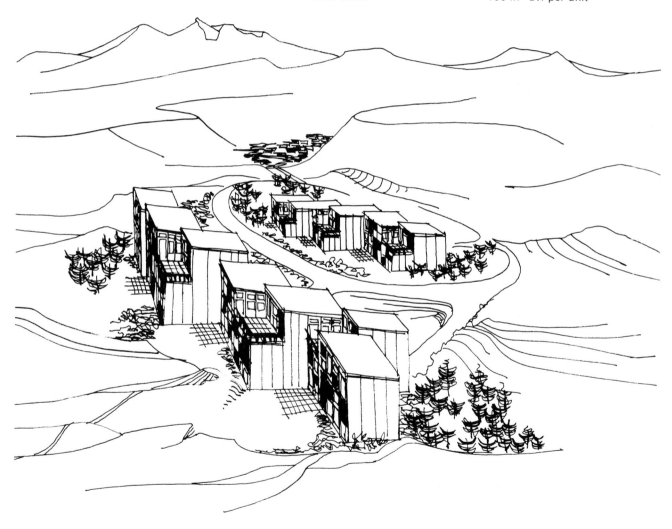

5.47
Odeillo houses

* Five similar houses constructed at Aramon, in co-operation with the French Electricity Authority, have water in metal tanks in lieu of the concrete mass. In this case concrete was chosen on cost grounds.

Location: Odeillo, Eastern Pyrenees, France
latitude: 43°N

Collector: as above in 1·8 m modular widths

Storage: 350 mm concrete walls (pointed various dark colours)*

Heating: as above
auxiliary: electric resistance heaters

Note: an internal temperature of 20°C is maintained with the solar input plus approximately 6000 kWh of electricity used for each unit per annum

5.24
Marseille design

Not a 'solar house', but a conventional building designed on climatological basis, which includes a responsive mechanism for the absorption or rejection of solar radiation

Designers: J L Izard and J P Long

Building: a terrace of 4 two-storey houses of five rooms each (Fig. 5.48)
floor area: 100 m² each

5.48
Marseille design: elevation and plans

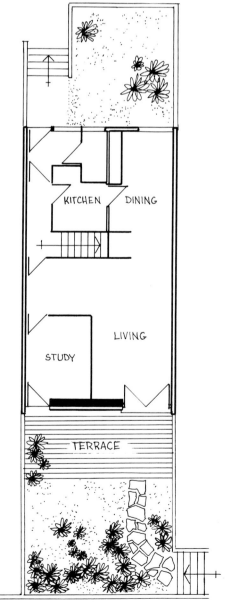

GROUND FLOOR

FIRST FLOOR

SOLAR WALL

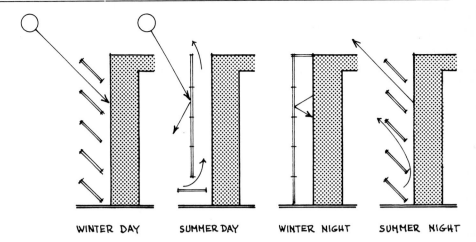

5.49
operating principles

WINTER DAY   SUMMER DAY   WINTER NIGHT   SUMMER NIGHT

Location: Marseille, France
         latitude: 43°N

The south wall of each unit is 40 m². Some 23 m² of this is a massive wall with an absorbent surface, covered with sets of aluminium louvres.

The method of operation is explained by Fig. 5.49.

5.25
Delaware house

Completed in 1973 for the Institute of Energy Conversion of the University of Delaware. Referred to as 'Solar One'

Designers: K W Böer, M Telkes et al.

Building: two-storey residence (Fig. 5.50)
           floor area: 132 m² (gr. fl. only)

5.50
Delaware house

Location: Newark, Delaware
          latitude: 39°N

Collector 1:   type: air + photoelectric
             position: south roof
             tilt: 45°
             area: 68 m² (in 24 panels)
             construction: 90 mm air space between cell array and 40 mm polyurethane backing
             glazing: single Plexiglas with Abcite coating

Collector 2:   type: air
             position: south wall bays
             tilt: 90° (vertical)
             area: 14 m² (in 6 panels)
             construction: black aluminium sheets
             glazing: as above

| Storage: | type: | latent heat of fusion |
| | location: | in basement |

A: $Na_2S_2O_3.5H_2O$ (sodium thiosulphate pentahydrate)

	mass:	3600 kg
	phase change:	at 49°C
	container:	sealed plastic trays (570 × 570 × 26 mm) stacked with 6 mm air spaces in a housing of 1·8 m cube, insulated
	heat capacity:	235 kWh

B: $Na_2SO_4.10H_2O$ (sodium sulphate decahydrate)

	mass:	630 kg
	phase change:	at 24°C
	container:	sealed plastic tubes of 32 mm dia, 1·8 m length in a 0·6 × 0·6 × 1·8 m insulated housing
	heat capacity:	41 kWh

C: mix of sodium chloride, sodium sulphate, ammonium chloride + borax

| | mass: | 1170 kg |
| | phase change: | at 13°C |

Heating:	collection circuits:	heated air to store
	1: roof —flow rate:	1·7 m³/s
	—velocity:	2·4 m/s
	—motor:	1·5 kW fan
	2: wall —flow rate:	0·6 m³/s
	—motor:	375 W fan
	distribution circuit:	warm air from store
	— motor:	375 W fan
	auxiliary 1:	heat pump from store B to A (24° to 49°C)
	— motor:	2 kW compressor
	— coeff. of performance:	3
	auxiliary 2:	electric heater
	— capacity:	9 kW
	— connection:	to store A

Cooling:	distribution circuit:	same as for heating
	disposal:	to store C
	heat pump (same):	from store C to outside air or to store B

| Electrical: | 68 m² CdS/Cu S cells (films) protected by Mylar efficiency: 5–7% (15–20 mA/cm² at 0·37 V) |
| | lead acid batteries of 180 amp-hour capacity inverted to produce A/C |

5.26
Atascadero house

Completed in 1973, referred to as 'Skytherm' house

| Designer: | Harold R Hay |

| Building: | single storey, seven-roomed residence (Fig. 5.51) |
| | floor area: | 108 m² + carport |

Location:	Atascadero, California
	latitude:	35°N
	altitude:	260 m

Collector:	type:	water container
	volume:	26·6 m³
	container:	8 bags of 0·1 mm polyethylene (11·4 × 1·2 × 0·26 m) on a black plastic sheet
	location:	on a ribbed steel deck flat roof which is also forming the ceiling
	cover:	9 panels of 50 mm polyurethane foam, sliding on five tracks

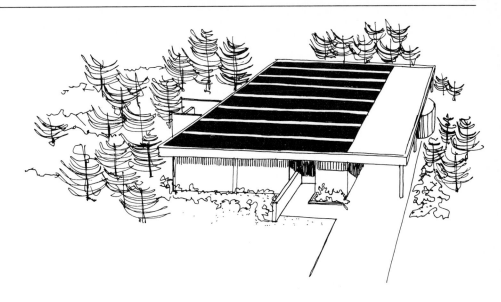

5.51
Atascadero house

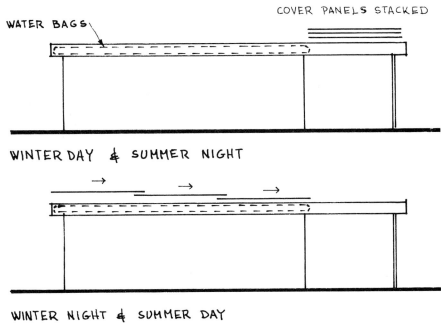

WATER BAGS

COVER PANELS STACKED

WINTER DAY & SUMMER NIGHT

WINTER NIGHT & SUMMER DAY

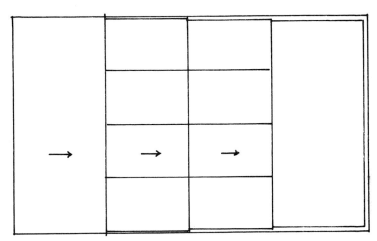

ROOF PLAN

5.52
operating principles

Heating: covers stacked over carport, water is heated and steel ceiling acts as radiator. Water bags are covered at night.

Cooling: water bags are exposed at night: radiant heat dissipation. Covered during day: cool ceiling gives radiant cooling effect (Fig. 5.52)

5.27
Milton-Keynes house

Not a 'solar house' but one unit in a terrace of a standard house design by the New Town Development Corporation, converted to partial solar heating. Scheduled for completion in 1974

Designers: S V Szokolay, P Atherton and D Hodges

Building: 1·5-storey, three-bedroom house (Figs. 5.53 and 5.54)
floor area: 84 m²
spec. heat loss rate: 250 W/degC
annual heat demand
(incl hot water):   13 500 kWh

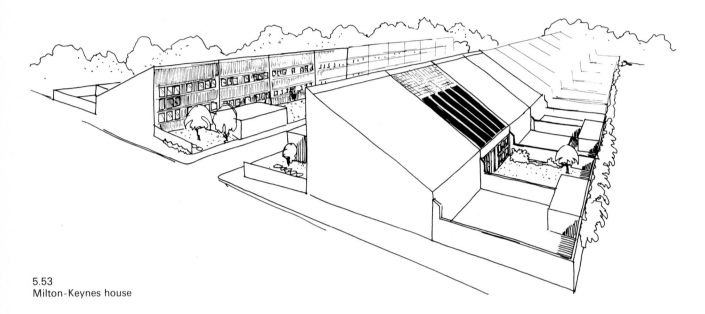

5.53
Milton-Keynes house

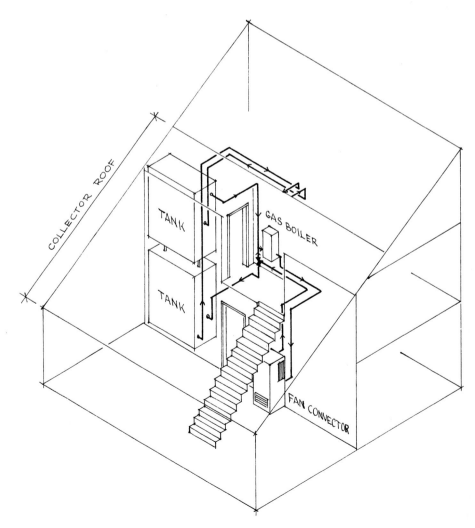

5.54
position of major components

Location:	Milton-Keynes, Buckinghamshire	
	latitude:	52°N
	degree-days re 18°C:	2600
	design t_o:	−1°C
	Jan. av. radiation:	490 Wh/m² day

Collector:	type:	water, +20% ethylene glycol
	position:	south roof
	tilt:	30°
	area:	43 m²
	construction:	

2·5 × 0·825 m Alcoa 'roll-bond' aluminium panels matt black anodised + 100 mm glass fibre insulation

	glazing:	single

Storage:	type:	water
	volume:	5 m³
	location:	in cupboard space on both floors
	container:	steel tanks
	heat capacity:	5·8 kWh/degC
	relative capacity:	116 litre/m² collector

Heating:	collection circuit:	heated water to store
	— flow rate:	200 l/h or 400 l/h
	distribution circuit:	warm water from store
	— flow rate:	200 l/h
	motors:	three pumps, 60 W each
	emitter:	central fan convector
	auxiliary:	gas boiler
	— capacity:	6·5 kW
	— connection:	in series
	operating temperature:	40°C (Fig. 5.55)

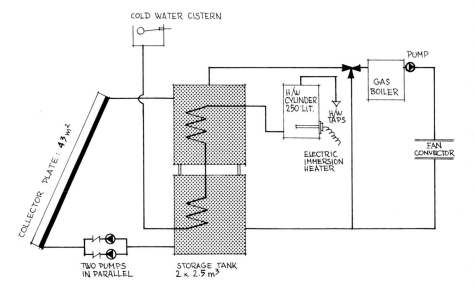

5.55
system diagram

Hot water:	flow-through type pre-heater	
	auxiliary:	electric immersion

Performance:	(computer simulation)	
	av. collect. efficiency:	30%
	Nov–Feb:	50%
	Jun–Aug:	23%
	auxiliary heat used:	5520 kWh
	solar contribution:	7980 kWh = 59%

5.28
Anglesey house

Designed for a private client, scheduled for completion in 1974

Designer:	S V Szokolay (solar installation only)

Building:	single storey residence (Fig. 5.56)	
	floor area:	82 m²
	spec. heat loss rate:	310 W/degC
	winter heat demand	
	(space heating only):	17 630 kWh

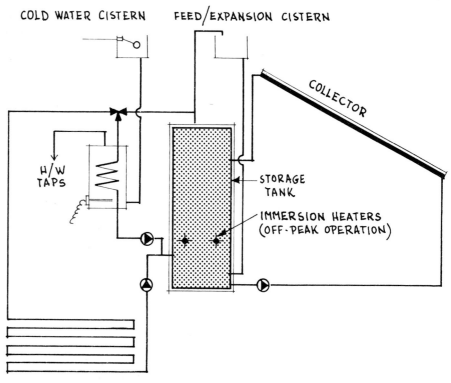

5.56
Anglesey house

Location:	Anglesey (Mallatraeth)	
	latitude:	53°N
	degree-days re 18°C:	2300
	design t_o:	0°C
Collector:	Type:	water+20% ethylene glycol
	area:	32 m²
	position:	south roof
	tilt:	30°
	construction:	
	2·15×0·825 m Alcoa 'roll-bond' aluminium panels matt black anodised+100 mm polystyrene insulation	
	glazing:	single
Storage:	type:	water
	(assisted by mass of concrete floor)	
	volume:	3 m³
	container:	steel tank
	location:	in cupboard space
	heat capacity:	3·5 kWh/degC
	relative capacity:	94 litre/m² collector
Heating:	collection circuit:	water to storage
	distribution circuit:	water from store

COLD WATER CISTERN   FEED/EXPANSION CISTERN

H/W TAPS

COLLECTOR

STORAGE TANK

IMMERSION HEATERS (OFF-PEAK OPERATION)

FLOOR WARMING COILS

5.57
system diagram

emitters: embedded copper coil floor warming

operating temperature: 26°C minimum
42°C maximum

auxiliary: off-peak electric
— capacity: 2 ×3 kW
— connection: in storage tank
(Fig. 5.57)

Hot water: in summer only, through indirect cylinder

Performance: (estimated)
solar contribution:
— to space heating: 7440 kWh =42%
— to water heating: 1300 kWh

5.29
Lincoln office

Proposed offices for the Massachusetts Audubon Society

Designer: Arthur D Little Inc

Building: two-storey office building (Figs. 5.58 and 5.59)
floor area: 750 m²
spec. heat loss rate: approx 1·2 kW/degC
max heating load: 41 kW
max air cond. load: 52 kW

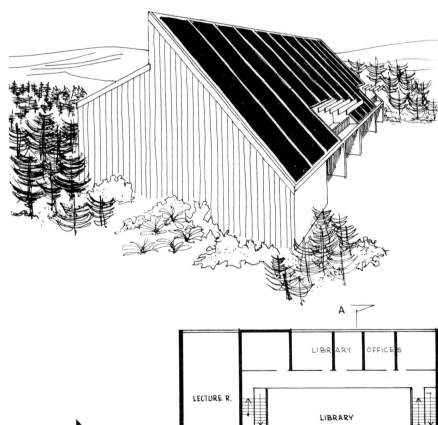

5.58
Lincoln office

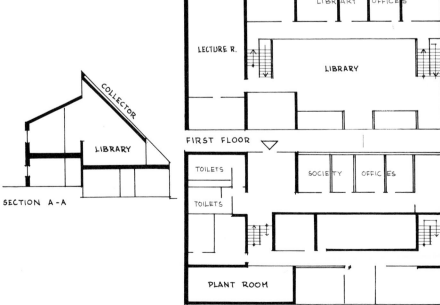

5.59
plans and section

Location:	Lincoln, Massachusetts	
	latitude:	42°N
	design t_0:	−14°C
Collector:	type:	water+ethylene glycol
	area:	330 m²
	position:	south roof
	tilt:	45°
	construction:	metal plate and tubes
	glazing:	double
Storage:	type:	water
	volume:	28 m³
Heating:	collection circuit:	water to storage
	distribution circuit:	water from storage
Cooling:	full air conditioning, using a 52 kW capacity lithium bromide/water absorption cooling unit powered by the solar collector	
Performance:	estimated solar contribution:	65–75%

5.30
Lake Padgett house

The first attempt by a firm of property developers (Covington Landmark Inc) to market a solar house. An experimental building was completed in April 1974 and the 'production model', 'Solar One' is scheduled for completion in 1974 (Fig. 5.60)

Design consultant:	E A Farber (University of Florida)	
Location:	Lake Padgett, Land o'Lakes, Florida	
	latitude:	29°N

5.60
Lake Padgett house

Collector:	type:	water
Storage:	type:	water
	volume:	25 m³
Heating:	ducted warm air from air handling unit in roof space	
Cooling:	chilled water to air handling unit from an absorption cooler powered by hot water from storage	
Hot water:	a 450-litre tank inside the main storage tank	
Note:	a swimming pool is coupled with the system and acts as a secondary heat sink, whilst being heated	

5.31
Avignon house

Private residence in an isolated location, designed to be independent of public utilities. To be completed in 1975

Designer:	Dominic Michaelis	
Location:	Avignon (Jucas), France	
	latitude:	44°N
Building:	single-storey residence (Fig. 5.61)	
	tubular steel space frame roof, deep verandah, lightweight well insulated wall panels	

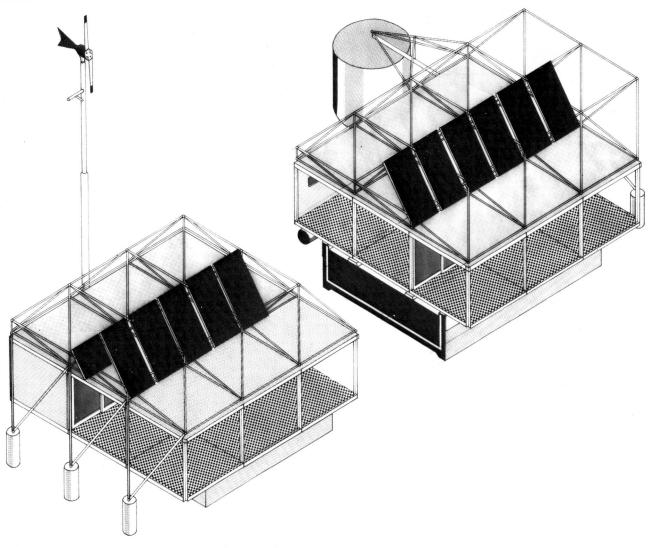

5.61
Avignon house

Collector 1:	type:	water
	area:	35 m²
		12 panels of 2.4×1.2 m
	tilt:	55°
	construction:	adhesive bonded aluminium panels
Collector 2:	type:	photoelectric cell array
	area:	1 m²
	tilt:	55°
	construction:	silicone cells (of Solar Power Corporation)
	expected output:	40 W normal
Storage:	type:	water
	volume:	150 m³
	temperature:	80°C expected max

Heating: 50 W fan to blow air over water surface (covered by a plastic film)

Note: electricity produced by the silicone cells is to drive the circulating pump

a wind generator is to supply domestic needs, and charge batteries; surplus electricity is used to heat water

inter-seasonal thermal storage is attempted

rainwater is collected in an external tank, filtered and used for domestic purposes

5.32
Chinguetti school

A solar-mechanical-pumping installation, designed to solve the water problem of this desert village. Completed in 1973

Designers: Alexadroff (M & Mme), Guennec & Girardier (based on the work of Prof. Masson of the University of Dakar)

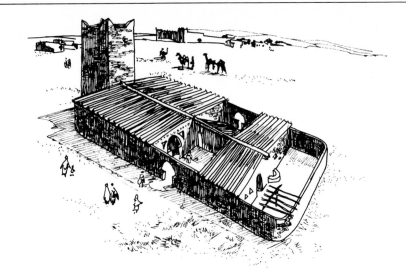

5.62
Chinguetti school

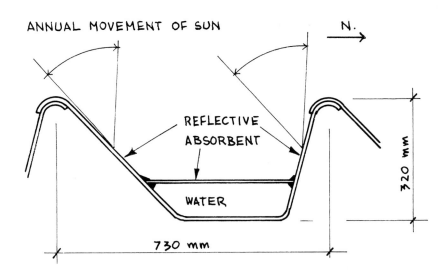

ANNUAL MOVEMENT OF SUN

N.

REFLECTIVE
ABSORBENT

WATER

320 mm

730 mm

ALUMINIUM ROOF TROUGHS ON E-W AXIS

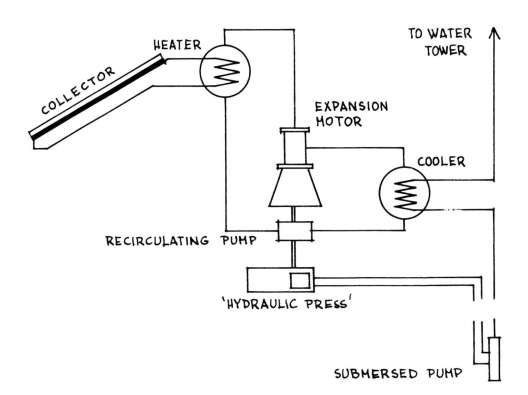

HEATER

COLLECTOR

TO WATER
TOWER

EXPANSION
MOTOR

COOLER

RECIRCULATING PUMP

'HYDRAULIC PRESS'

SUBMERSED PUMP

5.63
system diagram

Building: single-storey school, teacher's house and a water tower (Fig. 5.62)

Location: Chinguetti, Mauritania
latitude: 21°N

Collector: type: water
circulation: thermosyphon
position: roof
area: 88 m²
construction: aluminium troughs, semi-concentrating (Fig. 5.63)

Engine: low temperature expansion motor, operating temperature: 70°C

Performance: pumping height: 20 m
pumping rate: 8–10 m³/h
working time: 5–6 h/day

5.33
Timonium school

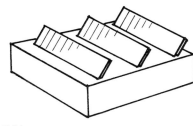

5.64
Timonium school

Aircraft Armaments Inc, with a grant from National Science Foundation, installed a solar heating system to one wing of the existing school. Completed in 1974

Building: single-storey primary school (Fig. 5.64)

Location: Timonium (nr Baltimore), Maryland
latitude: 39°N

Collector: type: water
area: 540 m²
tilt: 45°
position: three rows, free-standing on flat roof

Storage: type: water
volume: 57 m³
heat capacity: 4 days' demand

Note: collectors fitted with a honeycomb anti-convection device

5.34
Boston school

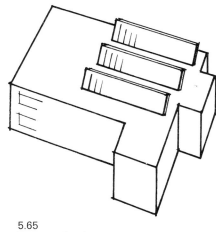

5.65
Boston school

General Electric Corporation (Space Division), with a grant from the National Science Foundation, installed a solar heating system to the existing building of Grover Cleveland Middle School. Completed early 1974

Building: two-storey school (Fig. 5.65)

Location: Boston (Dorchester) Massachusetts
latitude: 42°N

Collector: type: water + ethylene glycol
area: 426 m²
tilt: 45°
position: three rows, free-standing on flat roof
construction: 2·4 × 1·2 m roll-bond aluminium panels
glazing: double; 'Lexan' (polycarbonate plastic)

Storage: type: water
volume: 7·6 m³

Heating: air, heated in heat exchangers

Performance: solar contribution: expected 20%

5.35
Saginaw office

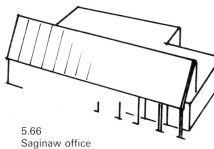

5.66
Saginaw office

An existing office block, housing a number of federal agencies, is to be fitted with a solar heating system. Scheduled for completion in 1975

Building: single storey office (Fig. 5.66)
floor area: 4700 m²

Location: Saginaw, Michigan
latitude: 43°N

Collector: type: water
area: 720 m²
tilt: 55°
position: on scaffolding over existing roof

The system is using a large water tank and heat pumps

5.36
Corrales house I

Built about 1970 in a semi-desert area, for the designer's own use. Named 'Zome-works Solar House'

Designer: Steve Baer

Building: single storey residence, consisting of interconnected polygonal units, with aluminium external finish (Fig. 5.67)

5.67
Corrales house I

Location:	Corrales (nr Albuquerque), New Mexico	
	latitude:	35°N
Collector:	type:	water, static mass
	position:	south wall
	tilt:	90° (vertical)
	container:	200 litre petrol drums in five rows on metal support-frame
	glazing:	double
Heating:	thermal storage effect only control:	
	— external:	fold-down shutter, reflective inside helps collection and acts as reflective insulator when closed
	— internal:	curtains reduce convection from drums; variable ventilation
Hot water:	separate solar water heater	

5.37
Corrales house II

The Davis house, built in 1974

Designer: Steve Baer

Building: single-storey residence (Fig. 5.68)

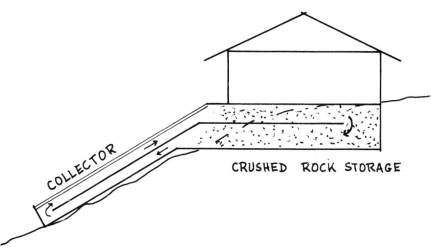

5.68
Corrales house II

Location:		Corrales (nr Albuquerque), New Mexico
	latitude:	35°N
Collector:	type:	air
	position:	outside the house on south-facing slope
Storage:	type:	rock
	container:	bin
	location:	under house

5.38
Shanghai house

Designed as a solar demonstration house for A N Wilson to be completed in 1975

Designer:		Burt, Hills and Associates
Building:		2·5-storey, 2-bedroom residence
	floor area:	132 m²
	(+garage and greenhouse)	
	(Fig. 5.69)	
Location:		Shanghai (nr Washington), West Virginia
	latitude:	39°N
Collector:	type:	water+ethylene glycol
	area:	56 m²
	position:	south roof
	tilt:	45°
	construction:	2·4 × 0·6 m, 1 mm aluminium panels with 6 mm aluminium tubes at 150 mm centres
	glazing:	2 layers Mylar film under a sheet of glass
Storage:	type:	water
	main tank:	
	— volume:	9·12 m³
	— construction:	concrete, integral with house basement
	— heat capacity:	2 days' demand
	supplementary tank	1·52 m³
Heating:		fan convector
Cooling:		in summer the supplementary tank is cooled and chilled water is circulated to the same fan convector

COLLECTOR

5.69
Shanghai house

5.39
Fort Collins house

A project of the Colorado State University Solar Energy Applications laboratory, supported by the University of Wisconsin and Honeywell Inc and funded by the National Science Foundation RANN (Research Applied to National Needs) programme. Completed in July 1974.

Designers:		G O G Löf, S Karaki and D S Ward
Building:		two-storey, residential type building to be used as a laboratory (Fig. 5.70)
	floor area:	284 m²
Location:		Fort Collins (nr Denver), Colorado
	latitude:	40°N
	altitude:	1560 m
	degree-days re 18°C:	3390
	design t_o:	−18°C
Collector:	type:	water
	position:	south roof
	tilt:	45°
	area:	73 m²
	construction:	roll-bond aluminium
	glazing:	double
Storage:	type:	water
	volume:	4·18 m³
	location:	on lower floor
Heating:	collection circuit:	water
	distribution circ:	water to heating coil in air duct

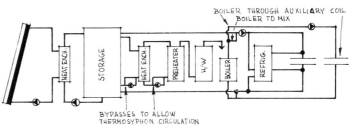

5.70
Fort Collins house

SYSTEM DIAGRAM

Cooling:	a modified absorption refrigerator of 10·5 kW rating (of the same make as used in the Brisbane house)
	refrigerant: water
	absorbent: lithium bromide
Auxiliary:	gas fired hot water boiler, feeding either to air coil for heating or to absorption cooler generator
Hot water:	pre-heating only
Performance:	expected solar contribution: 75%

5.40 Conclusion

This survey does not claim to be comprehensive, but it is certainly representative.

It is worth noting that the interest in solar houses seems to arise periodically and a burst of activity seems to be followed by a period of lull. Apart from the one pre-war example (5.2), we have four houses built in the immediate post-war period (1945–48). This is followed by a six-year gap. Sixteen examples are from the period 1954–60. The next decade saw a practically complete standstill, with a sudden burst of activity from 1970 onwards.

It may be argued that the lull of the 60's was caused by a general disappointment in the results previously achieved, especially in economic terms. Indeed, sceptics predict that the current burst of interest will be followed by a similar disappointment and relapse. It must however be obvious that the situation now is quite different to that 15 years ago. Then we were in the middle of the period of cheap energy, of boundless optimism and the general belief in growth. Since then we had fuel crises, which, although of a transitory nature, created an awareness of the vulnerability of our energy systems. The age of cheap energy is definitely over. The consequences of this will be examined in part 7.

There is also an increased environmental awareness and a growth of an ecological ideology. Pollution, resource depletion and the problems of growth are known and the consideration of these problems has started influencing people's behaviour and actions. The beginnings of a desire for living in sympathy with the global eco-system (rather than a short-sighted ruthless exploitation) are noticeable.

The technical problems of solar heating have largely been solved. Reviewing the above examples, three major types of solutions can be distinguished:

1   passive systems
2   thermal fluid (air or water) systems
3   systems using heat pumps

Each of these may be successful in its own terms.

1. At one end of the spectrum there is the Wallasey school (section 5.19) which largely relies on the thermal properties of the building fabric and offers very little control. This is the basis of the Rickmansworth house (5.11) (the heat pump is supplementary only). The Marseille design is marginally more sophisticated, with its adjustable louvres (5.24). The Chauvency-le-Château house (5.22) falls into the same category, the difference being that the latter has a glass skin, improving the collection but omitting the facility of shading. The Odeillo houses (5.23) are similar. Steve Baer's Corrales house (5.36) uses static water mass for heat storage instead of masonry, but the system is essentially similar. One variant of this system is the Atascadero house (5.26) where water on the roof surface is used as the heat store, with an adjustable cover. In principle it is nearest to the Marseille design.

2. Most houses surveyed have a thermal fluid system, which offers a far greater flexibility of control. Air is used as the collection fluid in 5.6, 5.7, 5.8, 5.9, 5.25 and 5.37. The large majority have water systems and practically all the recent examples (and those in the design stage) use water for collection and storage. Only three examples have attempted inter-seasonal storage, giving a total heat supply (5.2, 5.21 and 5.31), using huge water tanks for the storage of summer surplus heat (62 m³, 225 m³ and 150 m³ respectively). Of these probably only the Toronto design (5.21) would be fully successful.

Short of this, some form of auxiliary heat source must be used, as it would be uneconomical (even if possible) to use a collector large enough to supply the full demand at the worst part of the year.

3. The solar heating system can supply all the heat required only by using some form of heat pump arrangement. With this the heat pump may be considered as the auxiliary heat source. This arrangement is quite economical if the system is used in a dual role: heating in winter and cooling, air conditioning in the summer.

Some such systems use unglazed collectors as the heat source and heat sink (Tucson, 5.13 and Tokyo, 5.14). This will demand a practically continuous operation of the heat pump. Others use a glazed collector (Albuquerque, 5.10, Lincoln, 5.29, and Lake Padgett, 5.30), which gives satisfactory operation without the heat pump, using it only as an auxiliary unit for relatively short periods. Thomason (5.15 and 5.16) uses a glazed collector as his heat source and the unglazed roof surface as a heat sink, without relying on heat pumps.

A few have managed to avoid using electrically driven heat pump compressors by substituting a solar powered absorption cooler (eg Lincoln, 5.29, Brisbane, 5.20, and Fort Collins, 5.39, all three using a lithium bromide/water system).

Only two of the systems illustrated use photo-voltaic cells. Of these two, Avignon (5.31) uses it only to drive the circulating pumps; the Delaware house is the only one producing a significant amount of electricity. This is by far the most sophisticated, high technology system, involving not only the installation described in 5.25 but also a delicate and complex pneumatic control system.

It can be contrasted by Corrales I, which is a low technology, do-it-yourself, crude but successfully operating system. It seems that solar energy cannot be expropriated by any particular philosophy or attitude to life. The 'establishment' can make use of it just as much as those 'opting out'.

Part 6   Planning Implications

6.1
The problem

Individual buildings and the problem of their heating by solar energy has been discussed. Problems at a different level will arise if we start considering groups of solar buildings, housing rather than houses. With suburban densities the difficulties begin. The increase of density increases the problem.

Most of this section will be speculative. There is ample experience relating to the proliferation of solar water heaters but so far only the French are building groups of solar houses. Actual experience at this level is totally lacking. An attempt can however be made to foresee the problems that may arise.

The problems may be of two kinds:

1   physical, operational, ie how buildings are affected by their surroundings, by each other, problems of overshadowing, etc.

2   visual or aesthetic problems, ie how buildings affect the environment

6.2
Solar water heaters

In many warm-climate countries solar water heaters can be purchased in appliance shops, just as domestic refrigerators or washing machines. These are self-contained units, requiring only one electrical and two pipe connections. They can be mounted on roofs, as an afterthought, with no relation to the building (Fig. 6.1). One such device can be ugly, but if they proliferate on the roof of blocks of flats in a totally haphazard fashion, the townscape can be rather badly affected.

Some planning authorities now insist that these devices should be either integrated with the building or at least concealed from view.

Thau [1] has several suggestions:

1   collectors with integral horizontal tanks should be used in preference to vertical cylinders (Fig. 6.2)

2   units should be located in a line along the centre of the roof, to avoid visibility from below (Fig. 6.3)

3   they may be located near the northern edge of the roof, if provision is made for cover by a parapet and if view from above is likely, also by a partial roofing-over (Fig. 6.4)

4   they may be located along the southern edge (where they cannot operate if covered) only if the collectors can form a continuous strip running east-west (Fig. 6.5)

5   integration with the building, ie the use of collectors as building elements, is

6.1
a heater with no relation to the building

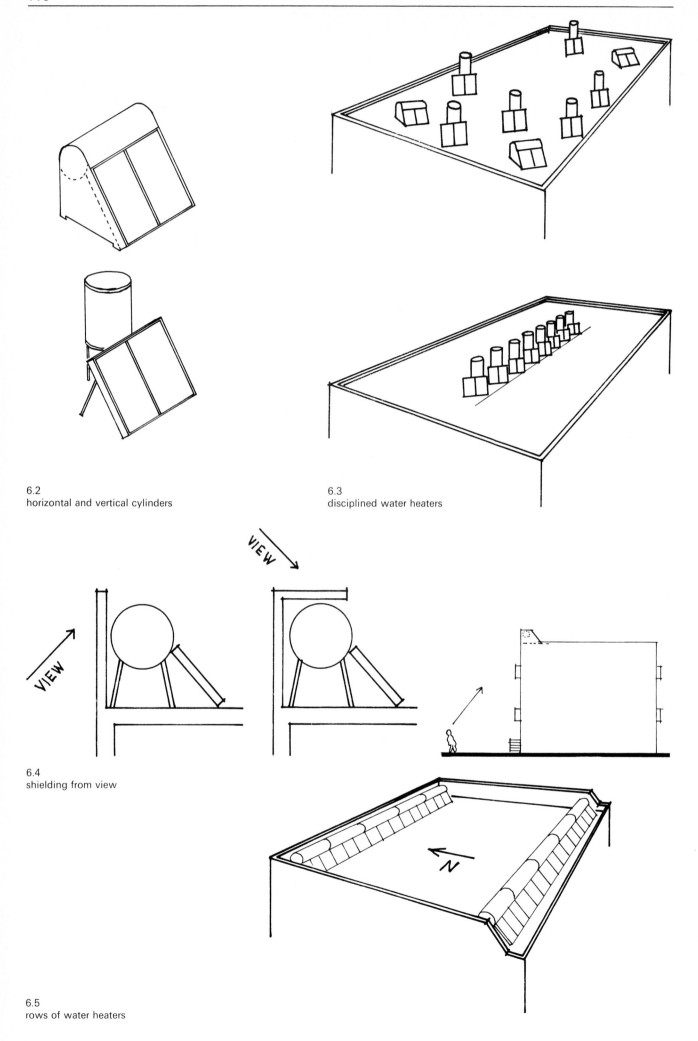

6.2
horizontal and vertical cylinders

6.3
disciplined water heaters

6.4
shielding from view

6.5
rows of water heaters

preferable to any of the above. They may become part of the roof, a balcony balustrade, a canopy over a window or any other form of shading device. (Fig. 6.6)

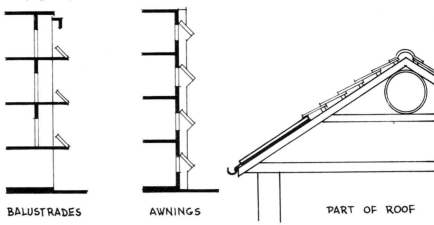

BALUSTRADES   AWNINGS   PART OF ROOF

6.6
integrated collector panels

6.3
Collectors for space heating: orientation

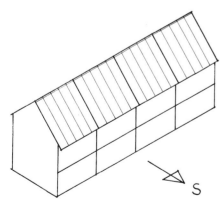

6.7
optimisation of orientation

The area of a collector for a water heating system is in the order of a few square metres. For space heating this will increase to several tens of square metres (typically 40–50 m² for a family dwelling unit). If the integration of the collector with the building fabric was desirable in the case of small water heaters, here it will be imperative. If so, then the collector orientation will determine the orientation of the building.

The optimum orientation is south, but several authors have suggested that 10° deviation east or west of south will not make any appreciable difference in the amount of energy collected. A computer simulation program constructed at the Polytechnic of Central London, using hourly measured radiation data from the Kew observatory (treating direct and diffuse components separately), produced a graph (Fig. 6.7) which shows that 50° deviation from the optimum orientation will give only about 10% reduction in collection. This may be considered as too much, particularly when the feasibility of an installation is marginal. It can however be suggested quite confidently that 30° east or west of south will give quite acceptable results, ie the reduction will be less than 2%. (One of the reasons for this is the high proportion of diffuse radiation, which is non-directional. This tolerance would be less in climates where the direct component of radiation is greater.)

The orientation of a detached house on a large site is fairly free. In a suburban situation however it is normally expected that the front of the building will be parallel with the street alignment. This requirement is even more positive with a row of terrace houses. If so, then this would dictate that the street should run east-west, at least if the collector is to be one side of a continuous double-pitched roof. (Fig. 6.8) If the gables face the street, then this should run north-south. (Fig. 6.9) In this case the bottom third (or so) of each south-facing slope cannot be used as a collector, as it will be significantly overshadowed.

There are several reasons why such straight-line arrangements need not be adhered to. Admittedly, these would normally produce the cheapest solutions, but also the most monotonous ones. A stepped layout may allow the use of large collector roofs, whatever the direction of the street. If, for example, the street was running NW to SE, the arrangement shown on Fig. 6.10 could be adopted. In this case the orientation of a sloping collector should be slightly east of south, to reduce overshadowing in the morning hours. Alternatively the area likely to be overshadowed (indicated dark on the sketch) may be left as an ordinary roof and not a collector.

6.8
terrace with collector roofs

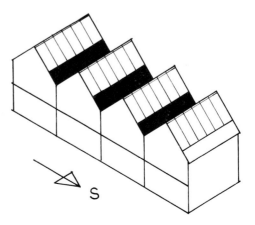

6.9
terrace with gable roofs

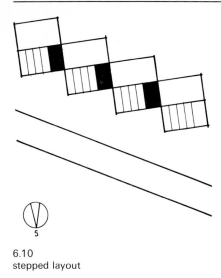

6.10
stepped layout

6.11
possible street directions

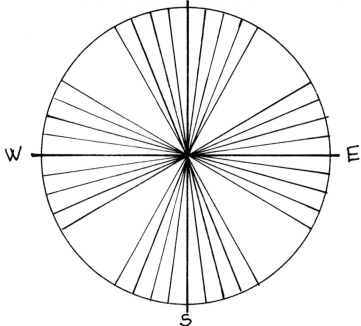

6.4
Collectors for space heating: tilt

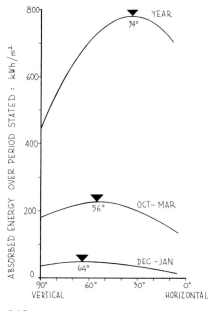

6.12
optimisation of tilt

6.5
Overshadowing

To sum it up: if the building form is rigid, linear, the street should preferably run east-west. The tolerance is ±30°, which gives a considerable freedom. A north-south street is also acceptable, again with ±30° tolerance. This means that 2/3 of all possible street directions are acceptable (Fig. 6.11). In any case, but particularly if the street direction is outside the above limits, the building form should be less rigid and some form of stepped plan will have to be adopted.

The literature abounds with recommendations for the optimum tilt of collector plates. Most frequently a tilt equal to the geographical latitude is recommended. Others suggest a tilt of latitude +10°, to give it a slight winter bias (as more heat is needed in winter). Still others recommend the use of vertical surfaces, partly because these would give a good performance in winter, partly to reduce the summer collection. The summer surplus of energy may otherwise become an embarrassment.

Where night-time radiant cooling is incorporated with the system, the collector-dissipator becomes a practically horizontal flat roof in almost all cases.

There is no one single answer to this question of tilt. It depends on local climatic factors. Most of the above recommendations take into account the direct radiation only. If diffuse only was considered, a horizontal plane would give the best results, as this would be exposed to the complete sky hemisphere. Thus the optimum tilt will be between that equal to the latitude and the horizontal, depending on and determined by the proportion of the diffuse component.

An optimisation program was constructed (see appendix 1), which gave the results shown in Fig. 6.12. This is valid for London, where 54% of the annual radiation is diffuse. This graph shows that to give a maximum all-year-round collection, a tilt of 34° should be adopted. To get the maximum in mid-winter (December–January), the tilt should be 64°. However, the advantages of the former in terms of collection quantities from early March to early October far outweigh the slight advantage of the steeper (64°) position in mid-winter.

The curves are not very sensitive to variations of the tilt. Practically any tilt between 30° and 70° can be justified by reasons other than the efficiency of collection. In building terms we may speak of a pitched roof or of an inclined wall, and a considerable freedom is available to the designer.

The decision regarding tilt will influence the collector's sensitivity to overshadowing. A collector of lesser tilt would be less sensitive to low level obstructions than a near vertical one. The stereographic sun-path diagram shows that the angle of obstruction which would start to affect the direct solar radiation depends on the direction of the obstructing object in relation to the orientation of the collector (Fig. 6.13). Towards the south any object may extend up to 15° altitude angle, without becoming an obstruction. In a direction 30° east or west of south (150° or 210°) this limit is 10°, but at the SE (135°) or SW (225°) an obstruction of about 3° would already cast a shadow on the collector in mid-winter. If we were to set this as the limit, it would be an unreasonable restriction. Intensities, when the sun is below about 10° altitude angle, are practically negligible.

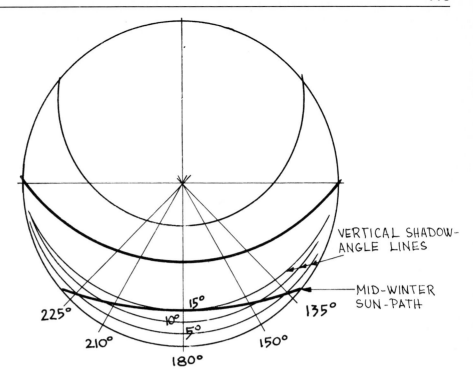

6.13
obstruction limits

We may generalise and say that obstructions in any direction up to 10°, and up to 15° towards the south will have no significant reducing effect on the amount of incident energy.

These angles must of course be measured from the bottom of the collector. (Fig. 6.14) The angle of 15° would mean that the height of any obstruction (above the level of the bottom of collector) must not be greater than about one-quarter of its distance.

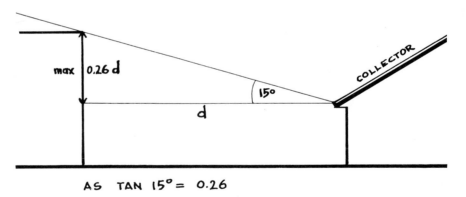

6.14
height and distance of obstruction

The decision regarding tilt will have to be made in knowledge of its visual consequences. A relatively low pitched roof is hardly visible from the ground level, the glazed collector surface will not be very conspicuous. The geometry of the situation is such, that any specular reflection from the collector glazing is unlikely to cause glare to passers-by. A vertical or a near vertical collector of a large area is likely to be the dominant feature of the building. It may have the advantage of better self-cleansing properties (dust is not likely to settle on it*) or better accessibility for cleaning, but it will be very conspicuous and reflections from the glass may cause glare.

6.6
Site survey

The graphic technique of establishing the extent of overshadowing by existing objects is quite simple.

a   take a point (P) at the centre of the lower edge of the proposed collector plate (Fig. 6.15)

b   draw a plan and a section at a vertical plane which goes through point P and lies in the direction of the collector's orientation

c   establish the vertical and horizontal angles subtended at the point by the obstruction, as shown on Fig. 6.15.

d   take a tracing of the shadow angle protractor and mark on it the angles, thus constructing the outline of the 'shading mask'

* Although it has been suggested that a slight tilt results in better wash-down by rain than a vertical surface.

e superimposing this shading mask over the sun-path diagram with the appropriate orientation, the duration of overshadowing can be read along the various date lines (Fig. 6.16)

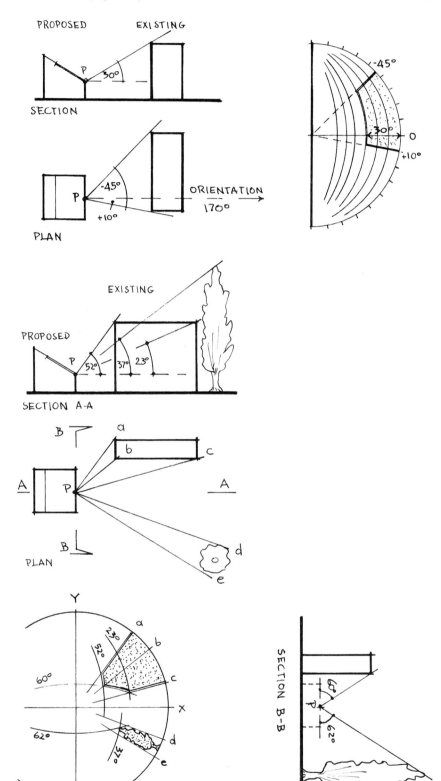

6.15
construction of shading mask

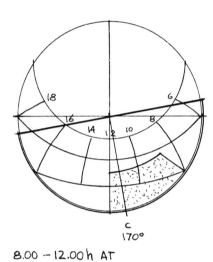

8.00 – 12.00 h AT
MID-WINTER
NONE AT EQUINOX

6.16
duration of overshadowing

6.17
a more complex shading mask

Fig. 6.17 shows the same process in a slightly more complicated situation. Note that for the construction of the shading mask each altitude angle must be established by pointing the centre line of the shadow angle protractor in the direction of the angle (in this case, for all angles of section A-A towards X, for the left side of section B-B towards Y and for the right side of section B-B towards Z).

The method can be used in reverse to establish the physical extent of the shadow cast on the collector at a particular time.

a superimpose the shadow angle protractor on the sun-path diagram with the appropriate orientation

b find the date and hour point on the sun-path diagram, read the horizontal and vertical shadow angles subtended by that time-point (Fig. 6.18)

c transfer these angles: vertical angle to the section and horizontal angle to the plan and draw projection lines from corners of the obstruction on to the collector plate. These will indicate the extent of the shadow cast

Such an assessment of the duration of shading and the extent of the shadow cast will help to determine whether there would be a significant reduction in the total solar energy received. It may help the decision whether to install a solar collector at all, and if so, what the most favourable position would be.

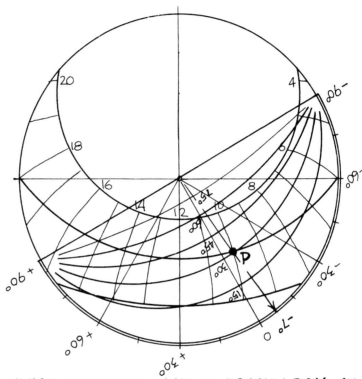

WHAT SHADOW IS CAST ON EQUINOX DAY AT 10.00h?
SEE POINT 'P'   (ORIENTATION 150° I.E. 30° E OF S)
ε = 30°
δ = -7°

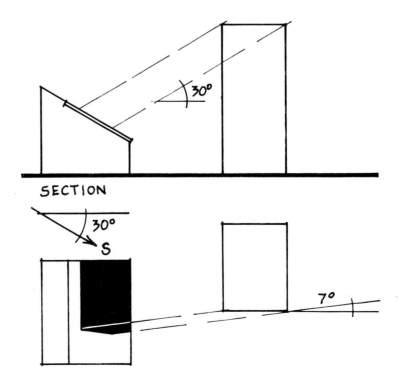

6.18
construction of shadow cast

6.7
Legal implications

Beyond surveying existing obstructions, the designer should assess the likelihood of future developments on neighbouring sites which may cause overshadowing at some later date. Existing town planning regulations may give some guidance (zoning, height restrictions) but no guarantee.

There is practically no legislation regarding the 'right of light' of buildings. Legal practice is based on precedent. In relation to new buildings there are several

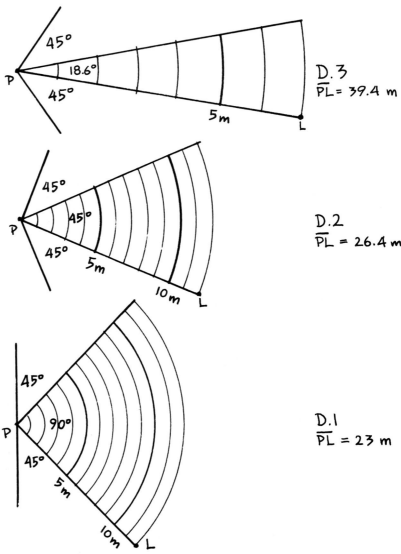

D.3
$\overline{PL}$ = 39.4 m

D.2
$\overline{PL}$ = 26.4 m

D.1
$\overline{PL}$ = 23 m

1:500   CAN BE REDRAWN TO ANY SCALE ON THE BASIS OF $\overline{PL}$ DIMENSIONS GIVEN.

HEIGHTS WITHIN SECTOR LIMITED.

10°   D.3

24°   SECTIONAL INTERPRETATION

        D.2

28°   D.1

6.19
the 'permissible height indicator'

advisory publications [2, 3] but it is up to the local authorities whether they enforce these. The most recent such set of recommendations are contained in the Department of the Environment booklet 'Sunlight and daylight, planning criteria and design of buildings' [4].

This takes March 1 as the critical date (identical sun-angles occur on October 15; thus the criteria given would ensure sunlight duration more than that specified, for 7·5 months of the year). The criterion is that on this date any building face oriented between azimuth 90° (east) and 270° (west) should be accessible to sunlight at all points 2 m above ground level for at least three hours. Times when the sun is below 10° altitude angle are not taken into account.

A set of tools, the 'permissible height indicators', can be used to measure whether this criterion is complied with. Fig. 6.19 shows one of these sets and explains its meaning, also giving a comparison with the requirement established above and shown in Fig. 6.14.

The purpose of these criteria is to protect the quality of the environment. It is not related to the utilisation of solar energy.

Society is not prepared for the utilisation of solar energy. Technically: most problems are solved. Economically: in some cases it is already feasible, in most cases it is on the limit of feasibility. Legally: it has not even been considered.

The building owner who invests in a solar heating system is not protected at all. A re-zoning of the area may permit higher buildings which may reduce the benefits of the installation. And he has no right for compensation. Even the growth of trees can interfere with the system and make him suffer a loss.

The problem should be approached from two directions:
town planning and
law of tort

The 'Sunlight and daylight' criteria should be extended to cater for the protection of solar collectors and should be made mandatory. It should work both ways: it should specify in what position and what level can solar collectors be installed with a reasonable chance of success. If installed outside such limits, they will have no protection. The 'right to sunlight' within such limits should be guaranteed. No new construction should be allowed to encroach on such limits.

Should such an encroachment occur, legal proceedings would be started and the defendant could be obliged to pay a continuous compensation equal to the value of energy lost on the collector due to his encroachment.

This may be a far cry, but at least it indicates the direction in which we should be moving.

6.8
A 'solar community'

A recent American study [5] suggested that a small community of about twenty dwelling units, with a centralised solar energy system would be more cost-effective than the independent individual house. With larger communities (in the order of thousands of dwellings) the cost of distribution would be disproportionately great.

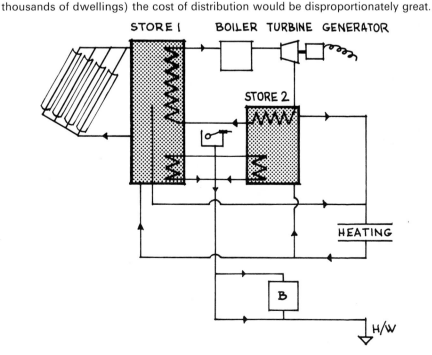

6.20
a community energy system

This is true if the energy system includes electricity production. If only thermal utilisation is proposed for water and space heating, it is unlikely that centralised systems would have any advantage over the individual house.

In this proposal focusing collectors are used. These are shallow parabolic troughs with an absorber tube, mounted on an equatorial axis (the axis being parallel to the earth's axis, ie running north-south, with a tilt equal to the latitude) with one dimensional (diurnal) tracking provided. These would produce temperatures in the order of 180–230°C (in store 1). This would drive a turbine and produce electricity (Fig. 6.20). Waste heat would be stored for heating purposes (in store 2).

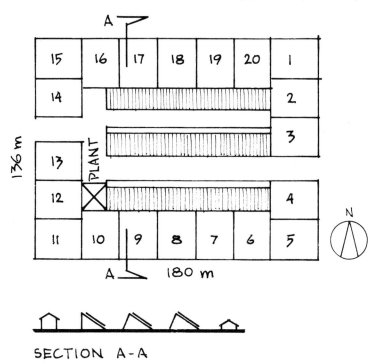

6.21
layout of a 'solar community'

SECTION A-A

Fig. 6.21 shows the layout of 20 plots with the collectors and plant located centrally. The site is 2·5 hectares, of which the plots occupy 1·5 ha (about 30 m × 25 m each); the plant and access takes up 1 ha. The total collector surface is some 2000 m².

Buildings on the plots along the south side would be restricted to single storey. The advantage of this layout is that the area where obstructions could overshadow the collectors is under the control of the users of the system, thus legal problems with third parties would be avoided.

References

1 Thau, A
Architectural and town planning aspects of domestic solar water heaters
in Arch Sc Rev. vol 16 No. 1, March 1973
2 British Standard Code of Practice
CP3: chapter 1 (B)—originally CP5: 1945)
Sunlight
3 *The Scottish Housing Handbook:*
1 Housing layout
rev ed 1958
4 Department of the Environment
Sunlight and daylight
(planning criteria and design of buildings)
HMSO 1971
5 McCulloch, W H, Lee, D O, Schimmel, W P
The solar community—energy for residential heating, cooling and electric power
Sandia Laboratories, Albuquerque 1974

Part 7 Economics and Prospects

7.1
Figure of merit

The question of collection efficiency has been considered in sections 2.4 and 2.5. The highest efficiency however does not necessarily mean the most economic system. High efficiencies can often only be achieved by a very expensive installation. Normally a balance must be found between capital expenditure and resultant savings in running costs.

A useful expression of this cost-effectiveness of an installation is the *figure of merit*. This can be taken as

$$FM = \frac{\text{value of energy saved by the installation in 10 years}}{\text{extra cost of installation over a conventional one}}$$

(some authors use a one year energy saving in the numerator).

It has been suggested that the system will be competitive if the figure of merit reaches a value of 1, ie if ten years' energy saving will equal the capital cost.

It must be remembered that the use of this figure of merit does not constitute a cost-benefit analysis. It simply gives a convenient comparative value. Meaningful cost-benefit analysis can only be carried out by using a 'discounted cash flow' technique, ie by converting future expenditure to its 'present worth'.

To obtain the value of energy saved, the annual heat requirement must first be established, then the amount of energy actually contributed by the solar collector is to be calculated and finally this must be given a monetary value.

7.2
Annual heat requirement

The annual heat requirement will depend on two factors:

a   a climatic parameter
b   a building parameter

The climate can be characterised by the *degree-day* concept. This can be described as the annual cumulative temperature deficit, ie the sum of the products of temperature differences and their duration. A reference level is agreed (say 18°C) as indoor temperature (t_i). The mean outdoor temperature is established for every day (t_o) and the temperature difference is taken as $t_i - t_o$. If eg the mean t_o is 2°C for three days, $3 \times (18-2) = 48$ degree-days are added to the sum. The exercise is repeated for all days of the year, whenever the t_o is less than the reference level.

This is purely a meteorological concept. Typical values are (re 18°C):

	degC days			degC days
Massachusetts	3300		North Scotland	3245
Colorado	3056		Midlands	2922
New Jersey	2830		London area	2616
Arizona	1020		Cornwall	2332

(18°C is taken as a reference level, assuming that incidental internal heat gains, such as from people and lighting, will ensure an internal temperature of 20°C.) These values may be converted to a reference level of 16°C by using a factor of 0·82.

If these values are multiplied by 24, we get the required climatic parameter in a more convenient form: the number of *degree-hours* (degC.h).

The building will be characterised by the *specific heat-loss rate* concept.

The heat loss rate through the building envelope is (cf section 4.8)

$$Q_c = (\Sigma A \times U)\Delta t$$

and the ventilation heat loss rate is (section 4.9)

$$Q_v = 0 \cdot 36 \times V \times N \times \Delta t$$

The total heat loss is the sum of the two:

$$Q = Q_c + Q_v = [(\Sigma A \times U) + 0 \cdot 36 \times V \times N]\Delta t$$

The heat loss rate per unit temperature difference is the *specific heat loss rate:*

$$\frac{Q}{\Delta t} = (\Sigma A \times U) + 0 \cdot 36 \times V \times N$$

dimensionally:

$$\frac{W}{degC} = m^2 \frac{W}{m^2\,degC} + \frac{Wh}{m^3\,degC} \times \frac{m^3}{h}$$

The value of this may be as low as 200 W/degC for a small well insulated house and as much as 1000 W/degC for a large loosely planned residence.

The product of the two parameters: the degree-hours and the specific heat loss rate will give the annual heating requirement

$$\frac{W}{degC} \times degC.h = Wh \qquad \text{(divided by 1000 to get kWh)}$$

To this value we may add the water heating requirement, if a combined system is considered. This will be the product of the daily hot water consumption, 365, the specific heat of water (1·16 Wh/litre × degC) and the increase from cold supply temperature to the required hot water temperature.

The cold water supply temperature is reasonably constant in the UK throughout the year at about 10°C. The hot water is normally supplied between 60° and 70°C, as this temperature is necessary for kitchen and laundry use (for bathroom use 45°C would be sufficient). Thus it is reasonable to take the required temperature increase as 65 − 10 = 55 degC.

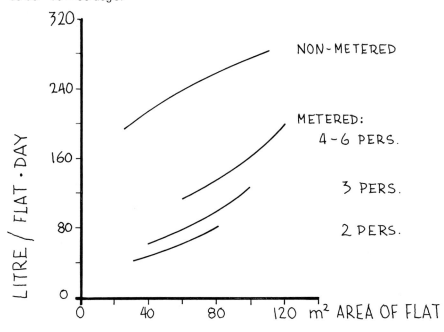

7.1
hot water consumption

For daily per capita hot water consumption the IHVE Guide [1] suggests the following values:

residences, low rental   70 litre/pers.day
          medium rental 115
          high rental   140

Rietschel and Raiss [2] suggest different values for metered and non-metered supplies (the latter would be paid for in a fixed sum included with the rent) — both as a function of the area of the apartment and not per capita. (Fig. 7.1)

7.3 Solar contribution

Unfortunately the supply of solar heat is out of phase with the heating demand. Much of the energy available in the summer will be wasted, as there is no simultaneous demand (in fact the collection system will be self-regulating to some extent: when the water temperature reaches the balance point, there will be no further collection, as the heat loss from the collector will equal the solar gain).

The amount of energy usefully collected will depend on seven parameters:

1 incident energy
2 optical loss through transparent cover
3 absorption properties of the receiving surface
4 heat transfer properties of the absorber (from surface to fluid, ie the 'plate efficiency', cf section 2.5)
5 thermal transmittance of transparent cover, which is a factor of heat loss
6 collection temperature, which in turn depends on
    a fluid flow rate
    b fluid temperature at entry to collector
7 external air temperature

The collection efficiency, ie the ratio of utilised energy to incident energy

$$\eta = \frac{Hg}{I . A . time} \qquad \text{(cf section 2.5)}$$

will be the result of all these parameters. Under favourable conditions it may be as high as 70% but it can be as low as 30%. Accurate prediction of the efficiency can only be achieved with a full simulation of the system's behaviour, taking into account all seven of the above parameters. This is clearly a task for the computer.

For the purposes of a crude estimate, with collection temperatures around 40°–50°C, it is reasonable to assume an average collection efficiency of 40% for the heating season.

The intensity of radiation constantly changes, but the cumulative total for a day or even for a month will provide a sufficient basis for an estimate. Table 7.1 gives average monthly totals measured on a horizontal plane (column 1), and calculated totals for a vertical south facing wall (column 2), as well as for an optimally tilted (34°) plane (column 3) (at Kew nr. London).

Table 7.1   Monthly amounts of radiation at Kew (nr. London)

	horizontal 1	south vertical 2	south 34° tilt 3
Jan	18·3 kWh/m²	30·3 kWh/m²	29·4 kWh/m²
Feb	30·9	47·3	51·6
Mar	60·6	61·8	81·8
Apr	111·9	75·9	137·1
May	123·2	57·2	133·2
Jun	150·4	53·8	155·7
Jul	140·4	53·6	142·1
Aug	125·7	69·1	141·1
Sep	85·9	75·2	111·2
Oct	47·6	62·8	72·8
Nov	23·7	41·2	40·5
Dec	14·4	22·6	22·2

It will be seen that the optimally tilted plane receives (on average for the winter eight months) about 1·5 times as much energy as the horizontal surface. If the values for a horizontal plane are available, these may be multiplied by the product of this ratio and the 40% assumed efficiency: $1·5 \times 0·4 = 0·6$. This would give the amount of collected energy for the period taken.

Even with manual calculations it will be necessary to work out at least monthly balances (both demand and collection) in order to establish the amount of effective solar contribution. The process is best illustrated by an example.

Using the above data and taking a house with a specific heat loss rate of 250 W/degC and a location with an annual number of 2800 degree-days (= 67 200 degC.h), having, say a 40 m² solar collector, we get the following monthly breakdown:

Table 7.2 Calculation of effective solar contribution

	deg C.h month A	space heat reqmt 0·25 × A kWh B	horiz total radiat'n kWh/m² C	unit collect'n C × 0·6 kWh/m² D	total collect'n D × 40 m² kWh E	solar contribution E but not more than B kWh F
Jan	10560	2640	18·3	11·0	440	440
Feb	9600	2400	30·9	18·5	740	740
Mar	9120	2280	60·6	36·4	1456	1456
Apr	6840	1710	111·9	67·2	2688	1710
May	4728	1182	123·2	73·9	2956	1182
June	—	—	150·4	90·2	3608	—
Jul	—	—	140·4	84·2	3368	—
Aug	—	—	125·7	75·4	3016	—
Sep	3096	774	85·9	51·6	2064	774
Oct	5352	1388	47·6	28·6	1144	1144
Nov	8064	2016	23·7	14·2	568	568
Dec	9840	2460	14·4	8·6	344	344
total	67200	16800	933	559·8	22392	8358

7.4 The value of heat

Having calculated the amount of solar contribution, its value in monetary terms should be established. The value of heat produced by any heating system can be taken as the product of

a   the fuel cost
b   its calorific value
c   total system efficiency

The running cost thus obtained can then be compared with the capital cost of the installation.

Fuel prices have increased quite radically in the last three years. Table 7.3 shows sales prices, calorific values, average system efficiencies and the value of unit (kWh) heat produced, using December 1970 and March 1974 prices.

Table 7.3 Fuel costs

	calorific value: kWh	average system effic'y	in December 1970		in March 1974	
			sales price	value per useful kWh	sales price	value per useful kWh
Coke	7·86/kg	60%	2·03p/kg	0·43p	2·68p/kg	0·57p
anthracite	8·02/kg	63%	2·09p/kg	0·41p	2·47p/kg	0·49p
house coal	7·09/kg	52%	1·45p/kg	0·39p	1·87p/kg	0·51p
fuel oil	10·60/lit	65%	2·04p/lit	0·30p	4·71p/lit	0·68p
paraffin oil	9·58/lit	70%	2·95p/lit	0·44p	6·16p/lit	0·92p
gas	1	56%	0·37p/kWh	0·66p	0·41p/kWh	0·73p
electricity,						
on peak	1	95%	0·78p/kWh	0·82p	0·95p/kWh	1·00p
off peak	1	90%	0·35p/kWh	0·39p	0·43p/kWh	0·48p

Accordingly, if in this example we consider substituting a solar heating system for (or incorporating it with) a gas fired central heating system, where the value of unit heat is 0·73 p/kWh, the annual saving achieved would be the 8358 kWh solar contribution (as calculated above for the 40 m² collector) times 0·73p = £61.

7.5 Figure of merit comparisons

For the purposes of a quick estimate it is reasonable to take the cost of collector as equal to the installation cost, for the following reasons:

a   the distribution system and most other components would be installed anyway, even with a conventional system

b   the collector roof would save some conventional roofing cost, which we can take as equal to the cost of a storage tank

If we take a collector cost of £20/m² (which is a feasible current market price), the extra cost of the solar heating system can be taken as $20 \times 40 = £800$, we get a figure of merit of

$$FM = \frac{610}{800} = 0.76$$

less than the suggested target value of 1. But if the cost of collector could be reduced to £15/m², ie the total cost to $15 \times 40 = £600$, we would get

$$FM = \frac{610}{600} = 1.02$$

which would make the system economically feasible.

It may be revealing to compare these FM values with that of direct conversion of solar energy to electricity.

The least expensive photo-voltaic cells cost about £300/m². With the best possible efficiency (15%) this would produce 225 kWh electrical energy in a year (in the south of England 1000 kWh/m² solar energy is received in a year on a horizontal plane or about 1500 kWh/m² on an optimally tilted plane, thus $1500 + 100 \times 15 = 225$).

At a cost of electricity of 1p/kWh, this would represent an annual value of £2·25, thus

$$FM = \frac{22.5}{300} = 0.075$$

Less than that for flat plate collectors by a factor of 10.

7.6
How much collector?

The analysis carried out above (section 7.3) shows that a significant amount of the heat collected will be wasted as there is no simultaneous demand. It is obvious that a small collector of only a few square metres would 'work' all year round; there would be a demand at all times for the heat it produces (if it is used for water heating as well as space heating). We may say that its *utilisation rate* would be 1. Its total contribution (or its share of the total heat demand) would be small. With increasing collector size the contribution will be greater, but the utilisation rate would be reduced.

The optimum collector size can be established by minimising the (say) 10-year total cost. For the introduction of the method some simplifications are used:

a   the collector cost is taken as a linear function (in fact the supply cost may be linearly proportionate to the area, but the unit installation cost would be reducing with larger areas). This is shown by the straight line 'A' on Fig. 7.2.

b   the annual net solar contribution with various areas of collector is calculated, continuing Table 7.2:

Table 7.4   Solar contribution for various collector sizes

all values in kWh	heating requ'mt B	unit collect' D	10 m² total	net	20 m² total	net	60 m² total	net	80 m² total	net
Jan	2640	11·0	110	110	220	220	660	660	880	880
Feb	2400	18·5	185	185	370	370	1100	1100	1480	1480
Mar	2280	36·4	364	364	728	728	2184	2184	2912	2280
Apr	1710	67·2	672	672	1344	1344	4032	1710	5376	1710
May	1182	73·9	739	739	1478	1182	4434	1182	5912	1182
Jun	—	90·2	902	—	1804	—	5412	—	7216	—
Jul	—	84·2	842	—	1684	—	5052	—	6736	—
Aug	—	75·4	754	—	1508	—	4524	—	6032	—
Sep	774	51·6	516	516	1032	774	3096	774	4128	774
Oct	1388	28·6	286	286	572	572	1716	1388	2288	1388
Nov	2016	14·2	142	142	284	284	852	852	1136	1136
Dec	2460	8·6	86	86	172	172	516	516	688	688
	16800	559·8	5598	3100	11196	5646	33588	10376	44784	11518
utilisation rate:				0·55		0·50		0·31		0·26

Values for 40 m² can be read from Table 7.2:

total: 22392
net: 8358   thus utilisation rate: 0·37

c   the net contribution values from the above are deducted from the total heating requirement, the differences are multiplied by 10 and by the unit heat cost (in

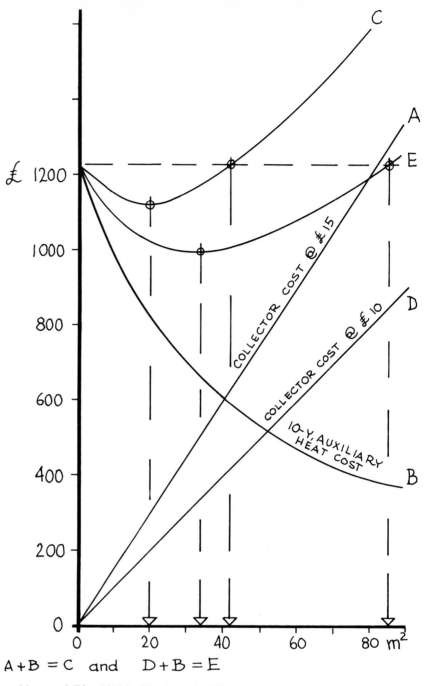

7.2
optimisation of collector area

$$A + B = C \quad \text{and} \quad D + B = E$$

this case 0·73 p/kWh); this gives the 10-year cost of auxiliary heating required, with various areas of solar collector. The results are plotted on the graph (curve 'B' of Fig. 7.2).

Table 7.5   Costs of auxiliary heating

collector area	auxiliary heat per year	×10 ×£0·0073
0	16800	£1226
10 m²	13700	£1000
20 m²	11154	£ 814
40 m²	8442	£ 616
60 m²	6424	£ 469
80 m²	5282	£ 386

d   if the two curves are added, the resultant (curve 'C' of Fig. 7.2) gives the 10-year total cost. The lowest point of this would indicate the optimum area of collector.

A point on this curve at the same level as the origin (on the Y axis) would indicate the maximum collector area, beyond which the 10-year cost would be more than that of a conventional system.

Using a collector cost of £15/m², the optimum collector area is shown to be 20 m² and the maximum about 42 m². It should however be noted that the optimum and particularly the maximum point is rather sensitive to variations in capital cost. A reduction of collector cost to £10/m² (line 'D') would produce the total cost curve 'E'—which indicates an optimum of 34 m² and a maximum of about 85 m².

7.7
Present worth

Particularly in these days of high interest rates the above method may give rather misleading results. We have in fact compared present expense with future savings.

One way to allow for the 'cost of money' is to use a discounting technique and establish the 'present worth' of future expenses (or savings). The capital expense will then be compared with this 'present worth'. An amortisation period must be assumed to start with, ie a period in which the extra investment should pay for itself.

The basic equation is

$$P = A \frac{(1+i)^y - 1}{i(1+i)^y}$$

where P = present worth
A = a uniform annual sum (eg annual heating cost)
y = number of years
i = interest rate (expressed as a decimal fraction, ie 12% as 0·12)

If we continue the above example, using an 18 year amortisation period and an interest rate of 12%, we get

$$P = A \frac{(1+0·12)^{18} - 1}{0·12(1+0·12)^{18}}$$
as $1·12^{18} = 7·69$

$$P = A \frac{6·69}{0·923} = A \times 7·25$$

Table 7.6 gives the results.

Table 7.6   Present worth of future heating costs

collector area	annual heating cost	P for 18 years
0	£122·60	£888·85
10 m²	100·00	725·00
20 m²	81·40	590·15
40 m²	61·60	446·60
60 m²	46·90	340·02
80 m²	38·60	279·85

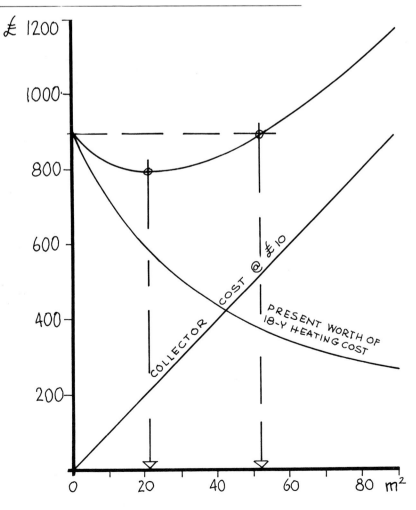

7.3
optimisation with discounting

The graph thus produced is shown in Fig. 7.3 (using the £10/m² collector cost). The optimum area is just over 20 m² and the maximum at 52 m² (as opposed to the 32 m² and 88 m² respectively, found above from a 10 year direct comparison).

The results would be identical with the 10-year direct comparison, when the expression

$$\frac{(1+i)^y - 1}{i(1+i)^y} = 10$$

eg with 8% interest for 21 years
or with 6% interest for 16 years

There is no allowance made in this method for inflation. With inflation future expenses (heating costs) would increase, thus the comparison would favour a larger investment, a larger collector area. It can indeed be argued that the current interest rates are high, because they include a compensation for inflation. It would be quite realistic to use an interest rate for such calculations equal to the difference between current interest rates and the annual rate of inflation. Eg if money is available at 14% and the annual rate of inflation is 8%, only $14 - 8 = 6\%$ should be used in calculating the present worth of future expenses.

Another factor which would influence such a cost-benefit analysis, is the future increase of fuel prices. Only a year before writing this, nobody would have dared to predict the increases experienced since. No one dares to predict today what the future increases will be, but everyone agrees that there will be increases well beyond the 'normal' inflation of all prices. We can only say qualitatively that such increases would further favour any present day capital expenditure leading to future fuel savings.

7.8
Cost-benefit studies

A recent report gives an account of a series of cost-benefit analyses carried out by Duffie and Beckman [3]. They take as *base cost* the initial cost common to both a conventional and a solar heating system and carry out their comparisons on the *costs above base*. They use the following values:

1 capital cost: cost of the solar heating system above base cost

2 annual cost of conventional system: fuel cost

3 annual cost of solar heating system: fuel cost in auxiliary heater(-s) plus cost of electricity used in the extra pumps

On this basis whole families of curves are produced, rather similar to the above, with ranges of various cost assumptions.

A much earlier study by Olgyay [4] used a discounting technique with a 20-year amortisation period and concluded giving a rule-of-thumb: a solar heating installation will be economically competitive if its capital cost does not exceed 2·3 times the cost of a conventional heating system.

7.9
Prospects

An American study [5] prognosticates that by the year 2020, 20% of the total US energy budget will be supplied by solar sources. For heating and cooling this proportion would be as high as 35%.

A New Zealand study [6] suggests that 24% of all electricity is used for water heating, of which at least half could be replaced by solar heaters.

No comparable study has been carried out in the UK. On the basis of numerous small scale experiments, some theoretical work and one experimental house project the following can be concluded (for the London region and the south of England):

1 a single glazed collector (eg 40 m²) of approximately half the floor area of the house (of 80 m²) would contribute about 60% of the annual space and water heating requirement

2 in terms of current fuel costs (April 74) the solar heating system will be economically competitive if the extra cost is not more than about £650. To achieve this, the collector should not cost more than £12/m²

3 present collector costs are over £20/m², but with mass-production techniques this could easily be reduced to £15/m². This price will be low enough to persuade clients with a long-term view, but not the average investor. He would want viability in terms of strict monetary values, would not spend more than £12/m²

4 a further 60% increase in fuel prices would make solar heating systems economically viable with this £15/m² collector cost

A good and inexpensive collector, a well organised solar heat industry offering a complete service and not just a product, could persuade many individuals to install a solar heating system. Nationally significant contribution can only be expected if both central and local government agencies are persuaded and pursue a vigorous programme of installations, primarily in new housing. Conversion of existing housing stock (in American terms: 'retro-fitting') can only be expected on a piecemeal basis.

The prospects of photoelectric devices very much depend on production costs. One typical module now available, which has a maximum output of about 1·5 W, can cost as little as £15 if purchased in large quantities. One supplier quoted recently the following prices:

for orders of   1–99   modules: £25 per module
               100–999   19
           over 1000   15

thus the corresponding cost per installed capacity would be £16·66, £12·66 and £10 (per watt) respectively. Current conventional generating plant costs can be taken as target figures:

for small plants, up to 200 W: £8·30 per W installed capacity
for medium plants, 1–2 kW: 2·00
for large plants: 0·20

This would suggest that for small plants, if ordered in large quantities, these cells are already almost competitive (£10 actual against £8·30 desirable), but even for 1–2 kW supply, let alone larger plants, they cannot be considered seriously (£10 actual against £2 desirable). For these, prices must be reduced by a factor of 5 before widespread application could be expected.

References

1 *IHVE Guide, 1970*
Institute of Heating and Ventilating Engineers
(London), 1971
2 Rietschel, H and Raiss, W
Die Heiz und Lüftungstechnik
14th ed
Springer, 1962
3 Duffie, J A and Beckman, W A
Modeling of solar heating and air conditioning
University of Wisconsin
Solar Energy Laboratory
Progress Report, January 1974
4 Olgyay, A
Design criteria of solar heated houses
paper S 93
'New Sources of Energy' Conf. Rome, 1961
UN 1964
5 NSF/NASA Solar Energy Panel
Solar energy as a national resource
National Science Foundation (USA), 1973
6 Vinze, S A
The place of solar water heating in the energy supply
paper E 68
'The sun in the service of mankind' Conf. Paris, 1973

Part 8   Solar Heat Industry

8.1   solar water heaters
8.2   swimming pool heating
8.3   photoelectric devices
8.4   building applications
8.5   refrigeration
8.6   costs

8.1
Solar water heaters

Solar water heaters were the first to catch manufacturers' imagination. These are small, relatively inexpensive items. Tangible products, which can be mass-produced and sold 'over the counter'. They fit the pattern of the domestic appliance market.

From the user's point of view a solar water heater is an attractive proposition, as it is relatively easy to fit it to an existing building and it gives a modest but definite saving. In most situations it is offered as a substitute to electric water heating, thus it is competing with the most expensive of fuels.

In Australia, for example, after some pioneering work the CSIRO (Commonwealth Scientific and Industrial Research Organisation, Melbourne, Victoria) published fairly detailed recommendations in 1959. Since then many small manufacturers started producing units which are rather similar to each other. Most of them use selective absorber surfaces, which give good results around 60–65°C and can ensure the supply of 70–90% of the annual water heating demand.

The largest of these firms is Beasley Industries (of Bolton Avenue, Devon Park, Adelaide, South Australia) who market the 'Solapak' collector panels and the 'Solatank' hot water cylinders. These can be arranged and installed in numerous combinations.

Braemar Engineering Co. (Bilsen Rd, Geebung, Queensland) have a similar range of products.

S W Hart & Co (of 112 Pilbara St, Welshpool, Western Australia) produce a neat collector panel with an integral horizontal cylinder, under the trade-name 'Solar-flo'. This can be connected to mains water supply, thus it does not need a feeder tank.

The product of Sola-Ray Appliances (of PO Box 75, Tuart Hill, Perth, Western Australia) is rather similar, with some minor improvements. They also produce swimming pool heaters and distillers.

Solarhot Water Systems (18 Fishers Reserve, Petersham, NSW), market swimming pool heaters as well as a hot water unit.

In Australia the annual production of solar collector panels exceeds 10 000 m².

Thermox Water Heaters Pty Ltd (15 Curtis Ave, Brisbane) have carried out several large installations in Papua-New Guinea, Fiji, other South Pacific islands, as well as in Australia.

In South Africa, Sadia Water Heaters Ltd (PO Box 43103, Industria, Transvaal) manufacture the 'Solapak' units under licence to Beasley Industries.

Solar Engineering Works Ltd (of PO Box 1760, Salisbury, Rhodesia) produce a wide range of units for commercial and domestic applications, including a special low cost unit for 'servants quarters' (!).

There are numerous small firms producing such heaters in Israel. The best known ones are:

Amcor Ltd (PO Box 2850, Tel Aviv), who produce a unit with an internally enamelled water tank, which it is claimed avoids corrosion and water deposit (furring) problems.

Mabua Co Ltd (of 80 Petach Tikva Rd, Tel Aviv) have a unit with built-in facilities for the removal of furring from the water channels.

In Europe, as far as it can be ascertained, the Italian firm: Industrie A Zanussi SpA (of via Montereale, Pordenone) are the only commercial producers of solar water heaters.

At the time of writing there are no commercially available water heaters in the USA, although several manufacturers are planning to market such units.

8.2
Swimming pool heating

After the pioneering work of Heywood in Britain there were some attempts to use flat plate collectors for water heating, but the only application which has achieved some degree of success, was the heating of swimming pools.

Lieutenant M Cribb had good results with his 160 m² heater to the Royal Navy swimming pool at Plymouth. The collectors were fabricated in the Navy workshops, thus it cannot yet be considered as 'industry'. After the installation of almost 50 m² flat plate collectors to a swimming pool in Hampshire, Solar Heat Limited (99 Middleton Hall Rd, Kings Norton, Birmingham 30) began producing and marketing a flat plate collector named 'Suntrap'. This is a copper plate with copper tubes, using 'Solarblak' coating, which is claimed to be a selective absorber.

Another firm, Drake & Fletcher (Maidstone, Kent), is now marketing an all-aluminium panel with rubber headers and have carried out several installations in the south of England.

These ventures are clearly aimed at the 'luxury end' of the market, but one can suggest three further reasons for the development of swimming pool heating and the stagnation of water heating development in the UK:

1   there are established solutions for water heating, efficient and reliable ones. It is hard to compete with these by a system which promises partial solution only. Swimming pools started proliferating in the south of England only in the last 15 years or so. The pool itself being a novelty, a novel method of heating is more readily acceptable.

2   the criteria in the case of swimming pools are not as precise and stringent as for water heating. All one expects is the extension of the swimming season and a slightly increased comfort.

3   large quantities of low grade heat are required for swimming pool heating, and —clearly—this is the field where flat plate collectors are the most efficient, even in the absence of bright sunshine.

During the last year, with the general increase in awareness of the energy problem, some of these firms started extending into the water heater market. Low Impact Technology Ltd (73 Molesworth St, Wadebridge, Cornwall) are for example marketing several British and imported heaters. A 'Solar Centre' shop has been opened recently in Chelsea (176 Ifield Rd, London, SW10) selling their own solar water heaters (under the trade name 'Sunstor') as well as several other products.

Swimming pool heating with solar devices is an established business in the USA, especially in the sunnier states, such as Florida or California.

The Energex Corporation (7988 Miramar Rd, San Diego) claim to be the first commercial producers of swimming pool heaters and they have now extended their activities into all areas of solar heat applications.

Alkins West (C-146, 5441 Paradise Rd, Las Vegas, Nevada) are marketing a 'high efficiency' solar energy pool heater, which they guarantee for ten years.

As large collector areas are involved and reduced efficiencies are acceptable, the unit cost of the collector is important. Fun & Frolic Inc. offer the cheapest collector panel under the trade name of 'Solarator'. Fafco (of Redwood City, Calif) are also in the market with a cheap collector very successful for swimming pool heating but unsuitable for higher temperatures.

8.3
Photoelectric devices

The development of photoelectric cells owes its success to the space programme. Several firms, originally producing such cell-arrays for space use, are now marketing them for terrestrial applications.

Spectrolab (12484 Gladstone Ave, Sylmar, Calif) produce a solar power system, which includes their LEC (light energy converter) cells, rechargeable batteries and a voltage regulator.

Solar Power Corporation (186 Forbes Rd, Braintree, Mass) are vigorously marketing a module, which contains five 55 mm diameter silicon cells, in a polycarbonate case, encapsulated in clear silicone rubber.

Several manufacturers apply this module for different purposes. In Britain, Joseph Lucas Ltd (Chandos Rd, London, NW10) market a battery charger assembly for use on caravans and boats. Unfortunately a small array of 450×345 mm, consisting of seven Solar Power Corporation modules, which produces maximum 10 W (under normal conditions 7 W) power, costs over £200.

AWA (Amalgamated Wireless of Australia) produce a radio telephone set and an automatic repeater station, both of which are powered by an array of the above mentioned modules. These are now accepted by the Post Office and can be linked into the normal telephone network.

Ferranti Ltd (Oldham, Lancashire) also produce a silicone solar cell and expect to market an array in the near future at a drastically reduced price.

8.4
Building applications

The use of flat plate collectors for the heating of buildings generally means that they would become a part of the building. The collector can no longer be considered as an appliance, or as a consumer product in its final use. It will become a building component. The installer comes between the manufacturer and the user.

From the point of view of the building industry—as we know it—there is an additional problem: the collector may be installed by the plumber, who does all the pipework, but it may also be part of the building envelope, normally the responsibility of the builder or of another sub-contractor, such as the roofer or the glazier. Some readjustment of trade divisions and responsibilities is necessary.

Few architects (and even fewer clients) would dare to take on the difficulty of obtaining all the components from various makers and organise all the trades necessary to execute the installation. Clearly, what is required is a complete service. One firm to deal with, rather than half a dozen, with an undivided and clearly defined responsibility for the full system.

As it stands now in Britain, industry is not yet organised. Practically all the components are available, but no composite assemblies or elements are offered and installation service is completely lacking. The few installations that exist or are now being constructed, experience difficulties which no routine job could tolerate or afford.

The purpose-made solar collectors are excessively expensive. Radiant ceiling panels or ordinary panel radiators may be used, but this means adaptation. The insulation and glass cover remain separate items, and problems to be solved.

In the USA the situation seems to be more developed. Several firms offer complete collector units as ready-made components. For example Sunworks Inc. (669 Boston Park Rd, Guilford, Connecticut) have two basic types: a flush and a surface-mounted module. They also offer a complete engineering service for all types of low-grade heat applications. PPG Industries (a subsidiary of Pittsburgh Glass) sell a collector which includes a roll-bond aluminium panel and a sealed double glazing unit.

The latest development is the offer of complete solar houses, such as that by Covington Landmark Inc (PO Box 308, Land o'Lakes, Florida), described in section 5.30, or by Energy Systems Inc (a division of Caster Development Corporation) of 643 Crest Drive, El Cajon, California.

Unfortunately there are also a number of small firms attempting to 'cash-in' on the 'energy crisis' panic, making performance predictions exaggerated out of all proportions. Such irresponsible salesmanship may do a serious disservice by discrediting the solar heat industry as a whole.

8.5
Refrigeration

In the field of thermal utilisation of solar energy probably refrigeration is the least developed.

There are only two successful experimental installations so far, using absorption refrigeration powered by solar collectors. Both are based on a cooler unit which was produced by Arkla-Servel (of Evansville, Indiana) but which has been discontinued. This was an Electrolux type unit of 10·5 kW (= 3 ton) rating, with gas heating, but relatively easily convertible to solar hot water operation. A few months ago the Colorado University group obtained (for the Fort Collins house) the last remaining unit in stock!

It is hoped that some manufacturers will realise the market potential and 'fill in the gap'.

An air conditioning installation based on such an absorption cooler will require roughly the same collector area as the heating installation of a similar building, ie approximately half the floor area. Thus the cost and availability of flat plate collectors is as important here as for space heating.

8.6 Costs

It is difficult to make cost comparisons, because of the wide variety of systems offered. In fact the only item for which such a cost comparison is feasible, is the flat plate collector. This is however the largest single cost item in most installations.

The following list is only a sample and by no means comprehensive. The first two products are suitable only for swimming pool heating and not for medium temperature (40–50°C) applications. (Addresses are given for firms which have not been mentioned before.)

	price per m²
Solarator	£ 3·50
Fafco	10·10
Warmswim (Drake & Fletcher, excl. glass)	24·00
Suntrap (Solar Heat Ltd, excl. glass)	25·00
PPG Industries	26·00
Sunworks Inc	30·80
Itek Corp. (Lexington, Mass)	30·80
Beasley Industries	44·00
Raypak Inc (Westlake Village, Calif)	52·80
Lockheed Corp (Palo Alto, Calif)	52·80
Inter Technology Corp (Box 340, Warrenton, Va)	55·00
Honeywell Corp (Minneapolis, Minn)	132·00

Some of these prices are difficult to understand, as very good collectors can today be produced, even in small quantities, for about £20/m² (including £4 for glazing). Probably due to small-scale production, the heavy development and overhead charges inflate the price.

On considering these prices, one is tempted to build one's own collector, buying only the absorber plate. The cost of several products suitable for use as an absorber can be compared:

A roll-bond aluminium panel, produced by Olin Brass Co (East Alton, Ill), sells for about £5/m², and the makers claim that this price could be halved with mass-production.

A standard radiant ceiling panel, made of steel sheet and a welded-on steel pipe coil, produced by British Steam Specialities, costs £9·55/m².

Ordinary central heating panel radiators produced by many firms cost about £9·20/m².

Alcoa aluminium roll-bond panels cost about £6/m² and a pattern suitable for solar collectors is likely to be marketed in the UK in the near future.

One American research group has set itself the task of designing and developing a complete collector unit which can be produced for £6·60/m² and be marketed for about £8/m². This they hope to achieve with an improved design, with the production technology in mind and with efficient production in large quantities.

Industry seems to be in the classical chicken-and-egg situation:

— mass production is only possible if there is a large market, but

— the market will expand only if the 'cost barrier' is broken, if there is a good, inexpensive product available.

Part 9

Design Guide

9.1
What device?

If the building designer has the idea that some form of solar device could be installed on the project being designed, the first question he faces is what type of device could be considered.

Electric power generation is out of the question at present, unless the situation is exceptional, ie if there is no public electricity supply available and the demand is only in the order of a few hundred watts (section 7.9).

Thermal utilisation is feasible in most situations. The first choice is whether the building should be designed as a collector, or some specific collector device should be installed. The former (ie the 'passive system') is definitely useful, but its success will very much depend on the user's correct actions at the appropriate times. This should be ensured by a clear set of instructions. It will rarely give the same performance as an 'active system' would, and certainly not the same degree of control. The strongest argument in its favour is its simplicity and relatively low cost.

With the second alternative the choice is between flat plate or focusing devices. The latter can only be considered if the climate is predominantly sunny. Even then the cost of the concentrator and the tracking system required will only be justified if there is a demand for temperatures near to or exceeding the boiling point. The use of flat plate collectors is on the other hand feasible even where the skies are predominantly overcast.

9.2
Is it feasible?

Technically, the collection of solar energy is possible in any situation on the earth. Feasibility is largely a question of point of view. The lowest efficiency system, with only a minimal contribution is an improvement and may be taken as feasible from the point of view of resource conservation. The speculative investor may not consider the most excellent system as feasible, even if it would pay for itself in five years. The average house owner may set the criterion of feasibility in terms of 10 or 15 years amortisation. Government agencies should (but rarely do) take a longer term view and consider a 20 or 25 year 'pay-back period' as satisfying their feasibility criteria.

Thus, before a feasibility study is started, the criteria (thus the stand-point) of the client should be clarified.

In the next step climatic factors may be considered. As a generalised rule it may be accepted that focusing devices are feasible if the annual total number of sunshine hours in the given location exceeds 2500 (ie about 6·8 hours/day average). This information is readily available from meteorological publications.

In a similarly generalised form it can be suggested that the feasibility of flat plate collectors is not governed by sunshine duration and their use may be feasible where the annual total amount of radiation, measured on a horizontal plane exceeds 800 kWh/m² (or approximately 2880 MJ/m² or 68 800 cal/m²). This information should be obtainable, if not from publications, from the nearest meteorological office.

This statement is based on a comparison of the expected heat collection with the cost of the collector:

As a rough guidance it can be assumed that an optimally tilted collector will receive about 1·5 times as much radiation as the horizontal and that a well designed system will achieve a 40% average efficiency; thus the heat collected will be

annual horizontal total × 0·6 (as 1·5 × 0·4 = 0·6)

With the 800 kWh/m² annual total we can expect 480 kWh/m² heat collection. How much of this will actually be utilised depends on the system and the energy use pattern. If we assume a utilisation rate of 0·5, one m² collector can save 240 kWh of heat which would otherwise have to be supplied by conventional sources. If the cost of such heat is, say 1 p/kWh and the collector cost is £24/m², we have achieved a figure of merit

$$FM = \frac{240 \times £0·01 \times 10 \text{ years}}{£24} = 1$$

(cf section 7.1)

If the figure of merit of 1 is accepted as the criterion of feasibility, one side of the equation is the 10-year fuel saving, the other side is the collector cost. Thus the criterion can be re-phrased to say that the system will be feasible if the collector cost (cf section 7.5 a and b) does not exceed the 10-year fuel saving, ie:

(annual total horizontal radiation × 0·3 × unit heat cost × 10) ⩾ collector unit cost

9.3
What system?

The choice is between air and water systems. This can only be decided in conjunction with the consideration of storage medium. Air systems (section 3.6) lend themselves for use with crushed rock or gravel heat stores. If the storage medium is to be water, then the use of water for collection fluid is preferable (section 3.4).

Both systems can be good, both have certain advantages and disadvantages. With air systems there is no freezing problem and no corrosion risk. However the collector, the ductwork and the storage will all be bulkier than with a water system. Air systems can normally be competitive in cost only if the building is designed around the system.

Water systems offer a greater flexibility in design and in operation and a higher degree of accuracy in controls. They can be readily coupled with conventional central heating and hot water systems (section 3.5). Provisions must be made to avoid the risk of freezing and corrosion (section 3.4 a, b, c). Any leakage of water may damage the fabric of the building—a risk which does not exist with air systems.

If the purposes of collection include air conditioning, ie the operation of an absorption cooler, then water is the only medium (or some other liquid) that can be used in the collection circuit.

9.4
What size collector?

A small collector will work usefully all-year-round, thus the capital investment is better utilised, but it will contribute very little in winter (section 7.6).

A simple rule-of-thumb is to use a collector of about half of the floor area. This may be followed in the sketch-design stage. When the design of the building has been completed, its thermal characteristics have been defined in terms of the specific heat loss rate, when the unit cost of the actual collector is known, a quick optimisation exercise can be carried out as described in section 7.6. The base data for this should include monthly total radiation values and the number of degree-days in each month. The technique can be refined and the results be made more reliable by using a discounted cash-flow method, introduced in section 7.7.

It must however be realised that both methods are based on certain efficiency assumptions (section 7.3) and that complete reliability could only be achieved by a full system simulation, in which the efficiency figure is produced as a conclusion and not used as a starting point. This however requires the use of fairly sophisticated computer programs, such as that described in appendix 2. Thus, it may be argued, unless the designer has access to some research group and can obtain such a computer analysis, there is not much point in carrying out an optimisation exercise, which is unlikely to be much more reliable than the rule of thumb quoted above.

9.5
What collector position?

The collector should obviously face the noon sun (south on the northern hemisphere and north on the southern hemisphere)—or nearly so. The question of tolerances has been discussed in section 6.3.

Regarding tilt, the simple rule-of-thumb suggests an angle equal to the geographical latitude. The validity of this has been discussed in section 6.4. A precise definition of the optimum tilt is possible if hourly data of diffuse and direct radiation (separately) is available and is used in a computer program such as that described in appendix 1.

This type of optimisation will show that there is a fair degree of tolerance. When eg the optimum is 34°, practically any tilt between 30° and 70° will be acceptable. Thus the architect has a considerable freedom. He may opt for a slightly inclined solar wall or for a solar roof, depending on his design concept influenced by factors other than solar. The exact tilt may then be determined by constructional detail considerations or by the availability of components.

9.6
What size storage?

The widely accepted rule-of-thumb suggests that for every m² collector area one should have a water storage volume of anything between 50 and 100 litres (or the equivalent in other forms of heat storage, giving a capacity of 58–116 Wh/degC).

The lesser volume will be more fully utilised, but by reaching a higher temperature more rapidly, it will result in a reduced collection efficiency. The larger volume may not be fully utilised, but would ensure a cooler temperature fluid returning to the collector, thus the collection efficiency would be increased.

The choice will depend on the desired collection temperature. If higher temperatures (around 50°C) are necessary, a smaller storage will be advisable. If lower temperatures (30–35°C) can be made use of, a larger storage will give better service.

Thermal properties of the building may also influence the choice. A massive, high thermal capacity building fabric can take over some of the storage function, thus the storage tank can be smaller. With a lightweight building, however well insulated, a larger storage tank may be necessary, even beyond the above quoted range, up to about 150 litres/m².

Finally, the climate pattern has a significant bearing on the desirable storage volume. Where there is a smooth transition from a cold winter to a hot summer, but there is not much day-to-day change, the storage capacity need not be more than a day's heat requirement. However, in a climate such as that of Britain, where there is no great difference between summer and winter, but almost the whole range of annual changes may occur within days, a larger storage is fully justified.

There is no point however in having a storage of more than 4 or 5 days' heat requirement, unless one decides to attempt an inter-seasonal storage. This would have to be larger by at least two orders of magnitude (eg 200 m³ as against 2 m³). The feasibility of this is yet to be proved.

9.7
What auxiliary heat supply?

Some auxiliary heat source is necessary, not just as a stand-by unit to be relied on in case of inclement weather, but as part of the normally operating system. Only in very favourable climates can the solar heat source be relied on exclusively in normal operation. It is suggested that in less sunny climates the solar source should normally be used as a preheater only, with the auxiliary heat source acting as a topping-up device. During favourable periods this topping up may not be necessary. The subject has been discussed in section 3.4 and a suggested circuit is shown in Fig. 3.11.

The choice of the actual heat source will very much depend on local factors. It may be chosen on the basis of conventionally applied criteria. It may be an ordinary gas fired boiler (if gas supply is available), an oil fired or solid fuel boiler (if storage facilities can be provided), or, failing these, it may even be electricity. If however off-peak electricity is used, the above suggestion will not be applicable. Then the heating element will have to be located in the storage tank and there is no possibility for a flow-through type topping-up arrangement. One solution for this problem is shown in Fig. 5.57.

9.8
What emitters?

The distribution system and the heat emitters may be any of the conventionally used types. The aim of the designer should be to make use of the lowest possible water (or air) temperature compatible with the heat delivery requirements. This would imply that either large surface radiant panels (eg floor or ceiling) or convectors with an increased heat transfer surface and air volume delivery will have to be used.

The discussion in section 3.4 may give some guidance, beyond this conventional criteria will be applicable.

9.9
What controls?

Where the designer (or someone intimately involved) is to be the user of an installation, manual controls may give a satisfactory operation. In all other cases at least the collection circuit should be controlled automatically. This can take three forms:

1 time-switch
2 solar switch
3 differential control

(A thermosyphon system would be self-regulating, but this would require an elevated tank, which is practicable only with small scale water heater units.)

The first is a rather crude device that would start the circulating pump a few hours after sunrise and switch off around sunset. It would ignore any weather variation. It would not only be wasteful in pumping power, but might in fact dissipate the heat already in store.

The second may be a photoelectric device or a thermopile, which would trigger off a switch when the radiation intensity reaches a pre-set level and switch off when it drops below this level.

A variant of this is the use of a solar cell array not just as a triggering device, but as the actual source of pumping power (cf section 5.31). Both these methods of control would follow weather variations, but would still ignore the state of the system, the temperature of the heat store.

The third system is the most sophisticated. It would use two sensors and a differential controller. A temperature sensor (a thermistor or a thermocouple) would be installed at the top of the collector plate near the flow outlet, another one at the bottom of the storage, near the return outlet. The differential controller would switch on when the reading of the first sensor exceeds that of the second one by a pre-set limit and switch off when the bottom temperature reaches the flow temperature.

Controls of the distribution circuit can be quite simple and the operation may be satisfactory with manual controls. If however a topping-up arrangement (such as the three-phase system shown in Fig. 3.11) is adopted, a manual control may be used to bring on the emitter (the circulating pump and the convector fan), but the phase selection should be made automatic.

Two sensors are to be used, one at the top of the storage and another in the return pipe past the fan-convector. When the first gives a high reading, the boiler is shut off. When it drops below a pre-set level, the burner is switched on. It acts as a simple thermostat. When however its reading drops below the reading of the second one, the differential controller will operate a motorised valve and the tank will be switched out of the circulation.

The manual switch operating the emitter may also be replaced by a time-switch, by a room thermostat or by a combination of both.

With all these controls the adjustability of settings is essential to allow for 'tuning' of the system during the commissioning period.

9.10
Developments

There is a tremendous scope for innovations both in system design and in constructional detailing. The field has not nearly been fully explored. The principle of such attempts should be to collect heat at the lowest useable temperature, and literally squeeze out every degree of temperature from the transfer fluid before it is allowed back into the collector.

Many designers got carried away in such attempts and produced overcomplicated systems. As in any design work, the product may go through a stage when it is highly complicated, but the most successful products will have matured into an elegant simplicity.

In constructional design, development is possible in several areas. The design of new emitters is not exactly an architectural task, but there is an opportunity of integrating them with the building fabric (eg floor or ceiling). The storage vessel could definitely be made to be part of the building, possibly even in a structural role.

The greatest scope is offered by the development of new collector types, not only efficient in operation, but also full-fledged building components. The metal parts carrying the thermal fluid could be shaped to act as structural members and span over longer distances. The insulating backing could be made to provide eg a ceiling finish. Producing cheap collectors (cf section 8.6) is one way of improving the economics of solar heating, but producing multi-purpose elements, thus obtaining savings in building is an equally valid approach.

Part 10   Progress in Some Countries

10.1
Japan

Already in 1973 more than 2·5 million solar water heaters were in use and much work had been done on other applications as well. There was however a qualitative jump in solar energy work in 1974, when the Ministry of International Trade and Industry established 'Project Sunshine'.

Funding for a 7-year period (up to 1980) of £127 (US$216) million has been approved and a total expenditure over 27 years (up to 2000) of £1483 ($2521) million is planned. [1] The administering agency has set up a strong systems engineering team and is directing the work of several specialised scientific teams.

The list of targets to be reached by 1980 includes the development of
—a 1 MW solar-thermal power generator system
—photovoltaic devices, 100 times less expensive than those currently available
—solar heating and cooling of buildings
—practical uses of solar furnaces.

Both the magnitude of the programme and its assured long range continuity are impressive.

Beyond this government-sponsored activity several large industrial concerns have invested much effort and large capital in various solar devices, such as absorption chillers and low temperature turbines.

10.2
United States of America

Following the NSF/NASA report (reference 5, p. 127) both a sub-panel of the Atomic Energy Commission [2] and the Solar Energy Task Force [3] have prepared reports and Congress has passed four Acts [4] establishing ERDA* (Energy Research and Development Administration) and a Federal Solar Energy Program. At about the same time the Department of Housing and Urban Development (HUD) and NASA have prepared a plan relating to residential buildings. [5] The NSF (National Science Foundation) has also submitted a report to both houses. [6]

In February 1975, an Interagency Task Force was established, chaired by ERDA. An interim report (ERDA-23) was produced in March 1975 and a revised version (ERDA-23A) in October 1975. [7]

The program has three main objectives:

*Officially established January 19, 1975.

1   demonstration of water and space heating and combined heating-cooling in commercial and residential applications, initially using *available systems* both in new buildings and retro-fitted into existing buildings

2 development of *new systems* in support of the above, initially using available components

3 research and *advanced systems* development.

There are five further 'sub-program areas':
— wind energy conversion
— bioconversion to fuels
— solar-thermal conversion
— photovoltaic conversion
— ocean thermal conversion.

It is expected that by the end of the century some 10% of the US energy needs will be supplied by solar sources.

The task of ERDA is to support the initiatives invited from industry, universities and other research organizations, to evaluate the results and to provide an information exchange. The initial funding of the program was $60 million and the proposal envisages an expenditure of $307 million over the 1975–9 quinquennium. The 1976 expenditure exceeded the original plan, it was approximately $180 million.

This is only the Federal Government's expenditure. Large sums have also been spent by industry and various research funds, often in co-operative projects. Typical of these are the following:

— Southern California Gas Co., in co-operation with the California Institute of Technology, is developing SAGE (solar assisted gas energy system) — a $1 million project, with about half of this amount contributed by NSF and ERDA

— Lockheed Palo Alto Research Laboratory, in co-operation with the City of Santa Clara: design, construction and installation of a solar heating and cooling system for a community centre (assisted by Solar Environmental Engineering Laboratory of Fort Collins) — grant support initiated by NSF and administered by ERDA.

10.3 Australia

The report prepared by the Australian Academy of Science in 1973 (reference 5, p. 28) was not nearly as seminal as its American counterpart. Some piecemeal developments, however, did take place and at the time of writing there are some hopes for an acceleration of developments.

Pioneering work has been done at the CSIRO (Commonwealth Scientific and Industrial Research Organisation) Division of Mechanical Engineering. In 1974 the CSIRO Solar Energy Studies Unit was created. Experimental work is continuing in the Mechanical Engineering Division. This Unit is primarily an information exchange, with a co-ordinative role, advising on policy matters, but also carrying out desk-type studies. The Unit administers the USA/Australia information exchange agreement signed in 1974.

The total government funding of solar research is some A$700 000 p.a. of which over $\frac{2}{3}$ is used by CSIRO and the remainder constitutes a few research grants to university-based projects. The largest grant from ARGC (Australian Research Grant Committee) has been received by the Queensland University group for fundamental research into photovoltaic effects in organic semi-conductors.

There is, however, a firmly established solar water heater industry, privately financed and prospering without any special incentives. The production and sales of such water heaters are rapidly increasing. Annual production figures (in terms of total collector area) are:

1971 — 3000 m²
1972 — 5000 m²
1973 — 8000 m²
1974 — 12 000 m²
1975 — 22 000 m²

In May 1976 the Australian Senate Standing Committee on National Resources began a series of public hearings on Solar Energy. It is expected that their report will be produced in the near future and the hope is that the government will act on the basis of their recommendations.

10.4 United Kingdom

The International Solar Energy Society (ISES), UK Section, prepared a report for the House of Commons Select Committee on Science and Technology (Energy Resources Sub-committee) in September 1974. The potential of solar energy in the UK was outlined and the need to expand research and development was stated. Following this the Society received a grant from a private foundation for the assessment of the role of solar energy in the UK. The panel convened has carried out a thorough study and produced a report in May 1976. [8] The Energy Technology Support Unit was established at Harwell in April 1974. They are preparing a report for the Department of Energy on the same subject.

The Department of Energy does not finance any research. The Department of the Environment was to chair an inter-departmental steering committee on solar energy research and was to administer the UK contribution to EEC work on solar heating and cooling of buildings, but recent political and economic developments have at least postponed any decisive action. The Building Research Establishment (BRE) of this Department is actively involved in solar energy work and plans to build three 'low energy' houses. The Housing Research section of the same department contributed to the funding of the Milton-Keynes project (para 5.27) and is co-operating with the Science Research Council in funding several university-based projects (eg the 'Autonomous House' project of Cambridge University).

10.5
France

Solar work has been well established since 1949, administered by the Centre National de la Recherche Scientifique (CNRS). The annual budget for research and development work is over £3 (US $5·1) million, but recently a doubling of this amount has been proposed.

The work centres on three areas:
—solar cells
—thermal applications in buildings
—solar-thermal power systems.

A 25 MW generating system is scheduled for completion by 1981, using optical concentration (the 'power-tower' principle), producing over 400°C steam. [9]

Independently of the above, the Electricité de France (EDF) has financed the construction of 10 solar heated experimental houses.

10.6
German Federal Republic

The Ministry of Science and Technology has an Energy Research Directorate (Projektleitung Energieforschung) dealing with non-nuclear energy research projects. The annual budget for solar work is some £3 (US $5·1) million.

Considering that work in this country only started in 1974, remarkable progress has been made. Projects are jointly funded with large industrial firms, mostly on a 50-50 basis.

The main problems are seen to be in the technology/economics field. Most of the projects relate to water and space heating [10], but there is some work on latent heat storage, on photovoltaic systems and on thermo-mechanical electricity generation at the 10 kW scale.

10.7
EEC

The Council of Ministers of the European Economic Community approved a solar research and development plan in August 1975. Details of the agreement are still under discussion. The proposed annual budget is around £1·8 (US $3·06) million, and projects envisaged include:

—flat plate collector development for heating and cooling
—solar power generation at the 1 to 10 kW scale as well as at the 1 MW scale (the latter including hydrogen production)
—photovoltaic device improvement
—fundamental research into photochemistry, photo-electrochemistry and photo-biology.

10.8
NATO

The North Atlantic Treaty Organization, Committee for the Challenges of Modern Science (CCMS) set up a pilot study on solar energy in 1974, relating to solar heating and cooling systems in buildings. The study is administered by ERDA (in Washington, DC).

The first meeting of the representatives of 17 countries was held at Odeillo (France) in October 1974. A Zero Energy House Panel was established at that meeting under the aegis of NATO/CCMS and it has organized further meetings in Copenhagen, Aachen and Palo Alto. The main purpose of this panel is information exchange and dissemination, thus indirect influencing of government decision-making bodies.

10.9
Israel

Israel has no special agency for solar energy work, but some 20% of all governmental research funding is devoted to solar energy projects. This amounts to approximately £2·5 (US $4·25) million per annum. [11] Most of these funds are given to industrial research laboratories, less than 10% (£250 000, US $425 000) to university-based projects. Thus the majority of the work is treated as confidential and reaches the general public only when a project is marketed. For example, a solar powered Rankin-cycle turbine unit has been developed, is now marketed and has already been supplied to many countries in sizes from 600 W to 5 kW.

There are over 50 firms producing solar water heaters and the 'market penetration' is deeper than in any other country. A great impetus was given to the use of such heaters by the abolition of the 15% purchase tax last year. Pressure is now mounting for the introduction of penalty rates for electricity used in water heating. The townscape of Tel Aviv is dominated by solar water heating. Haifa City Council put a ban on the use of such heaters for aesthetic reasons, but this decision has recently been modified to allow such units if they are integrated with the building design.

Research and development work is going on at six universities and one other research centre relating to

—central hot water systems for blocks of flats
—combined space heating and cooling systems
—solar ponds
—photo-biological glycerol production

as well as refinement and perfection of technology in many existing applications.

The largest single user of solar energy is the potash industry at the Dead Sea. In some 130 km² evaporating ponds they use the equivalent of 10 million tonnes of oil (when the total oil use of Israel is some 7 million tonnes).

10.10 Canada

The Brace Research Institute of McGill University has been active in solar energy research since 1959. In 1975 this Institute has prepared a report for the Ministry of Science and Technology, surveying the work done in Canada and abroad, assessing the potential of solar energy in Canada and made proposals for a 20-year research, development and education programme.

10.11 Egypt

The National Research Centre, Solar Energy Laboratory, has produced a study assessing the feasibility of dual solar and wind power systems for the northern coastal region of Egypt. The design of such a system is in progress.

10.12 India

There is no information available on any government funding, but interesting and meticulous work is done at several centres, mostly at the theoretical and analytical level, but also in practical applications, especially related to agriculture, eg pumping (vapour-lift pumps), waste disposal, distillation. The All-India Solar Energy Working Group has an important co-ordinative and information exchange role.

10.13 Italy

There is no information regarding any government support, but quite significant work is done in at least two universities. The work of Professor Francia at Genoa University is mentioned in section 11.6. At Naples University there is an active solar group. Their most interesting work relates to selectively emitting surfaces which can be used for cooling by radiation to space (even in daytime) in certain narrow wave-bands for which the atmosphere is transparent.

10.14 Mexico

The 'Subsecretaria de mejoramento del ambiente' (sub-secretariat of environmental improvement) of the government has initiated a solar energy programme, mainly concerned with solar-mechanical devices for the purposes of pumping water in arid areas. Most of this work is derived from Alexandroff, Guennec and Girardier (cf section 5.32). Some original theoretical work on solar refrigeration is done at Mexico City University.

References

1 *Japan's Sunshine Project*
Sunshine Project Promotion Headquarters, Agency of Industrial Science and Technology, 3–1 Kosumigaseki 1-chome Chiyoda-Ku, Tokyo, 1974

2 Atomic Energy Commission
Subpanel IX report (chairman Dixy Lee Roy)
Dept. of Commerce, National Technical Info. Center,
WASH 1281-9
September 1973

3 *Solar energy task force report*
submitted to the Federal Energy Administration for
Project Independence Blueprint
Government Printing Office, 4118-00012
November 1974

4 a Solar Heating and Cooling Demonstration Act, 1974
       September 3, 1974, PL 93–409
   b Energy Reorganisation Act, 1974
       October 11, 1974, PL 93–438
   c Solar Energy Research, Development and Demonstration Act, 1974
       October 26, 1974, PL 93–473
   d Federal Nonnuclear Energy Research and Development Act, 1974
       December 31, 1974, PL 93–557

5 *Program plan, solar heating and cooling demonstration program (residential dwellings)*
Dept. of Housing and Urban Development—National Aeronautics and Space Administration
SCH-1001
December 1974

6 *National solar energy programs*
submitted by National Science Foundation to US Senate Com. on Labour and Public Welfare and to House of Repr. Com. on Science and Technology
December 1974

7 *National program for solar heating and cooling*
(residential and commercial applications)
ERDA, Division of Solar Energy
October 1975

8 *Solar Energy, a UK Assessment*
report by the UK ISES panel
May 1976

9 *Research and development in the domain of energy*
Le Progrès Scientifique (DGRST)
March/April 1975

10 Friednich, F J
Solar energy projects for heating of buildings in Germany
paper at NATO/CCMS conf. Palo Alto
August 1975

11 private communication from Prof. D Wolf
Ben Gurion University
December 1976

Part 11

Development in Applications

11.1
Flat plate collectors

Developments in flat plate collectors are taking place in two directions:

1   producing inexpensive low-temperature collectors to operate at 50–60°C for domestic hot water production

2   creating higher performance collectors, to achieve acceptable efficiencies at around 100°C temperatures.

The first direction is pursued by using materials other than copper (such as steel, aluminium, plastics) and fabrication techniques less labour intensive than the tube-grid and plate absorbers. Welded pressed steel, roll-bond aluminium and moulded plastic collectors are coming to the market, with varying degrees of success.

The second direction is followed by

a   maximising radiation transmission of the transparent cover, eg, using low-iron glass or coating the glass with a non-reflective surfacing. The Kalwall Corp. of Manchester N.H. is reported [1] to have a fibreglass sheet ('Sun-lite') with transmittance equal to or better than glass, and a long infra-red transmittance not greater than glass. It has a lower thermal conductivity and higher impact resistance. Non-reflective coatings can also be applied to it.

b   minimising heat losses from the collector. Back and edge losses are reduced by better insulation. In frontal losses the radiant and convective components must be tackled separately.

Radiant losses can be reduced by better selective surfaces (cf section 2.4, p. 17). One aim is to produce a higher a/e ratio, another is to achieve durability under high temperature exposure. Many institutions are working on such surfaces, typical of which is the work of the CSIRO Division of Mineral Chemistry. [2] They have reported the development of a chrome black surface with an a/e ratio of almost 10, without any degradation when exposed to 150°C temperature for six months.

Convection losses could be reduced by using a vacuum between the absorber surface and the transparent cover. Attempts, such as that mentioned in section 2.6 were not entirely successful, as the junction between dissimilar materials will never be perfectly airtight, thus the vacuum will be lost after a certain (varying) period. G Francia in Italy has developed a collector in which convection losses are reduced by a plastic honeycomb structure placed between the absorber surface and the glass cover. The AAI Corporation has installed collectors using the same principle (but an aluminium honeycomb) at the Timonium school project (cf section 5.33).

11.2
Evacuated glass tubes

Another line of development was followed by several glass companies. Philips in Germany, Corning Glass and Owens-Illinois in the USA have been working on evacuated glass tube collectors for over two years. The all-glass enclosure gives a permanent airtight seal. The structural properties of the tubular shape allow the use of quite thin wall thicknesses. The technology is very similar to the manufacture of fluorescent tubular lamps. None of them are marketed yet, but some have been installed in experimental houses.

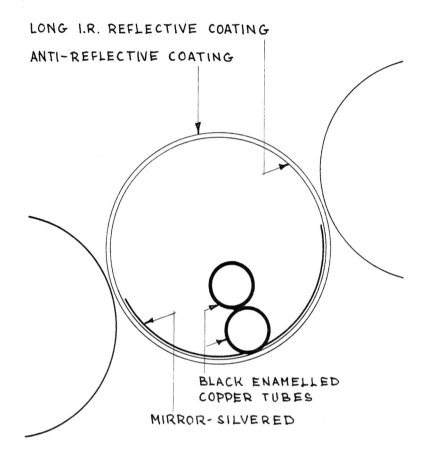

LONG I.R. REFLECTIVE COATING

ANTI-REFLECTIVE COATING

BLACK ENAMELLED COPPER TUBES

MIRROR-SILVERED

11.1
the Philips evacuated glass tube collector

Fig. 11.1 shows the Philips system. As the vacuum practically eliminates convection losses and the selective black absorber surface radically reduces radiant losses, very high performance can be expected even at high temperatures. Fig. 11.2 gives a set of performance curves for the Corning tubular collector.

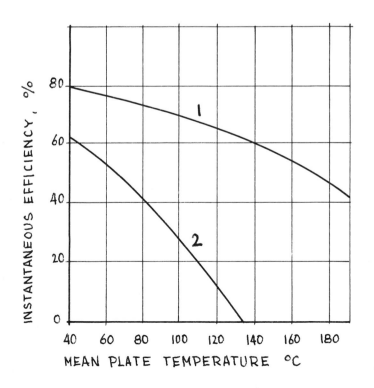

11.2
performance curves (at $t_0 = 10\,°C$, $I_{pt} = 946\,W/m^2$)
1 : Corning evacuated glass tube
2 : a typical double glazed, non-selective flat plate

11.3
Low concentration collectors

The generator temperature required by the lithium bromide/water absorption cooler units is in the region of 80–90 °C. Even the best flat plate collectors are very inefficient at such temperatures. The operation of Rankine cycle engines and turbines normally requires over 100 °C source temperatures. This is beyond the reach of flat plate collectors.

As such temperatures can only be produced with clear sky radiation, the advantage of flat plates, namely that they respond to diffuse radiation, is lost. Two simple low concentration devices are now on the market. One is the Northrup unit, patented, but actually developed from a design by Szulmayer. [3] This employs a linear Fresnel lens (or diffraction strip) of a plastic material: initially rigid PVC, later an acrylic plastic. A 300 mm wide strip can focus on a 13 mm diameter tube, giving a concentration rate of 24. As the practical efficiency is around 0·75, an actual concentration rate of about 18 is achieved. In the Northrup unit up to six troughs are mounted on a frame, tilted to the latitude angle and have a one-directional tracking mechanism (east to west at 15° per hour).

A similar frame and tracking mechanism is used by the Swiss firm Liebi LNC Ltd. Their collectors are, however, simple parabolic troughs with a mirror surface and a line focus. 6 to 18 such troughs (each having an aperture of 1.83 m²) can be mounted on a frame and share the tracking motor with a push-rod connection. The breakthrough in this case is claimed to be an extremely durable and easily cleaned glass mirror surface, applied to a steel sheet trough, using a method developed from porcelain enamelling technology.

The performance curves of both types are shown in Fig. 11.3 (based on sales brochure data).

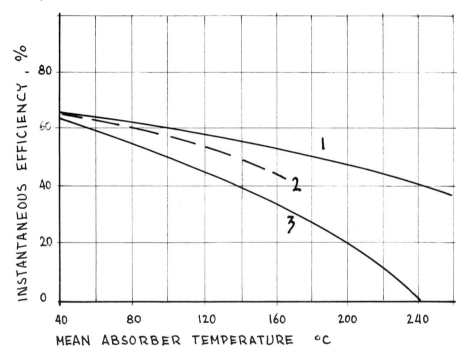

11.3
performance curves (at $t_o = 20$ °C, $I_{pt} = 1000$ W/m²)
1: Liebi LNC, with selective coating
2: Northrup trough, selective coating
3: Liebi LNC, non-selective black

Szulmayer has recently developed an ingenious system which requires no tracking. [4] The trough, with reflective interior and covered by a linear Fresnel lens is mounted horizontally on an east-west axis. At normal incidence the radiation is focused on the absorber tube by the lens, directly. At oblique incidence the specially shaped inside surface reflects the convergent rays, as it were, completing the focusing. A single strip concentrator has an angular tolerance of 15 to 20° (Fig. 11.4/a) and a 3-strip arrangement will accept radiation between 10° and 80° solar altitude (Fig. 11.4/b). Test results showed that stagnation temperatures up to 200 °C can be reached even with less than 500 W/m² horizontal intensities. Efficient heat extraction up to 150°C is feasible.

11.4
Air conditioning

The lithium bromide/water absorption chiller unit used in the Brisbane solar house (cf section 5.20), originally produced by Arkla-Servel, has been further developed and modified for low temperature (solar) applications. A Japanese company (Yazaki Corporation) is marketing such units with a 4.6 kW or a 7 kW cooling capacity. A generator temperature as low as 75 °C can keep these units operating. The coefficient of performance can reach 0.68 at 77 °C generator temperature, with a 24 °C sink temperature. These units have been installed in experimental houses as far afield as Switzerland, USA and Australia. Recently the same firm has introduced a 26 kW unit.

The York division of Borg-Warner Corporation has produced a unit with 175 kW

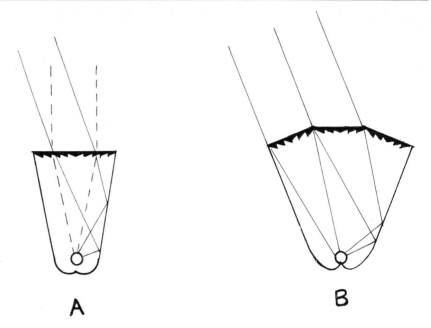

A   B

11.4
'W'-trough reflectors with Fresnel lens
a: single strip trough
b: 3-strip trough

cooling capacity. One of these has been installed at the Timonium school (cf section 5.33) and is operating with reasonable success.

An Israeli firm (Tadiram) has also developed a small chiller unit, which is to be marketed in the near future.

The University of Florida group [5] has developed several ammonia/water chiller units, which can operate with generator temperatures between 57° and 82°C, with corresponding coefficients of performance of 0·3 and 0·6. These can produce below freezing point temperatures. Previous studies suggested that operation would cease when the generator temperature drops to 74°C.

An entirely different line of approach is followed by the James Cook University group at Townsville. Their system is based on evaporative cooling with subsequent humidity control by adsorbent materials. These adsorbents (eg silica gel) are then reactivated by using solar heated air to expel the adsorbed moisture.

The 'Solar-MEC' system of the Chicago Gas Developments Corporation works on similar principles. It uses one heat transfer wheel (or rotary heat exchanger, cf Fig. 4.1) and another wheel with an adsorbent impregnated paper honeycomb. The two wheels provide a transfer of heat and moisture between the air intake and exhaust ducts in the desired direction. The unit using 1 m diameter wheels has a cooling capacity of 10 kW. A cooling capacity of 175 kW is expected from a planned unit using 4·2 m diameter wheels.

11.5
Medium temperature thermo-mechanical systems (90–130°C)

The principles of expansion engines have been discussed in section 3.8. In the last two years several developed units appeared on the market.

The Rankine-cycle engine developed by MBB (Messerschmitt-Bölkow-Blohm) GMBH in Munich uses a Freon organic fluid (R-114: $C_2Cl_2F_4$). The system consists of 700 m² flat plate collectors, producing heat at 95°C. The Freon is heated to 90°C. At this temperature it becomes a super-heated vapour and drives a Linde screw-turbine, developing 16 kW mechanical power. This can drive a generator of 10 kW electrical output.

The Ormat turbine, produced in Israel, works on similar principles, but in this case the working fluid is mono-chloro-benzine. The evaporator temperature needs to be around 110°C. Several sizes are available between 3 and 8 kW mechanical output. They are successfully used in several countries for water pumping and electricity generation.

The Japanese IHI (Ishikawajima-Harima Heavy Industries) Company has been producing Rankine cycle turbines in sizes up to 3800 kW, for industrial waste heat recovery, for the last 10 years. Recently they have introduced two small units (25 and 50 kW) for lower generator temperature applications, using Freon (R-11) as the working fluid, which they refer to as SHAORCE (solar heat actuated organic Rankine cycle engine).

As discussed in section 2.8 (heat pumps) the efficiency of these units is a function of the evaporator-to-condenser (source-to-sink) temperature differential. Fig. 11.5 gives the efficiency curves for the IHI unit, using R-11 as the working fluid. From this it

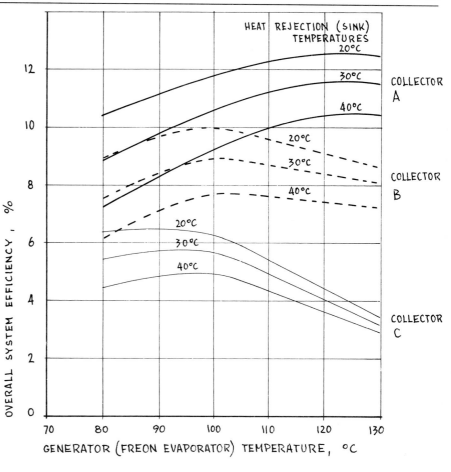

11.5
overall system efficiency:
$$\eta = \frac{\text{work output}}{\text{incident radiation}}$$
of the IHI Rankin-cycle turbine with working fluid R-11 and three different flat plate collectors

will be obvious that the success of thermo-mechanical devices depends partly on the availability of a low temperature sink, partly on the development of high efficiency collectors, giving a reasonable performance at over 100 °C temperatures.

Barber [6] has reported on the performance of a Rankine cycle engine, with a working fluid of R-113, which is used to drive a compression cooling system, with a working fluid of R-12. The following efficiencies were obtained:

power cycle	0·09
refrigeration cycle	7·40
mechanical	0·75
overall COP	0·50

11.6
High temperature thermo-mechanical systems (>200 °C)

The 'power tower' principle has been introduced in section 3.13 (Fig. 3.38/a). Since then Professor Francia has built a small-scale prototype system in Genoa. It consists of 121 circular individually adjustable mirrors, covering an area of 7 m × 7 m. The boiler suspended above this field can produce up to 21 kg/h of superheated steam at 500 °C driving a turbine. [7]

A small American company (Solar Physics Corp. of Lakeside, Calif) is marketing a domestic scale system, with an electrical output of 5 kW. This consists of 600 mirrors of 300 × 300 mm size. Each has its own aiming and tracking mechanism, which is claimed to cost not more than $1 each (!). Fig. 11.6 shows the diagrammatic layout of this system.

Honeywell Inc. has constructed a small-scale prototype at Minneapolis and is in an advanced stage of planning a 10 MW generation station. This will consist of 74 000 individually steerable mirrors of 3 m × 6 m each, covering an area of about 2·5 km², with the boiler-collector located on top of a 500 m high reinforced concrete tower. Steam over 500 °C will be generated to drive the turbines.

The Solar Energy Laboratory of the University of Houston (Texas) is working on the design and construction of a similar system. A 2 km × 2·4 km area is to have a 40% coverage by 4 to 6 m flat silvered mirrors, focusing on to a boiler on top of a 300 m tower. The steam turbine driven generator output could reach 100 MW. Chemical energy transmission and storage is investigated.

11.7
Photoelectric systems

In single crystal silicon cells the development over the last few years has been mainly in gearing up for larger-scale production, increasing sales and consequent reduction in prices. In section 7.9 the price of £16·66 (US$28.35) per watt peak output has been

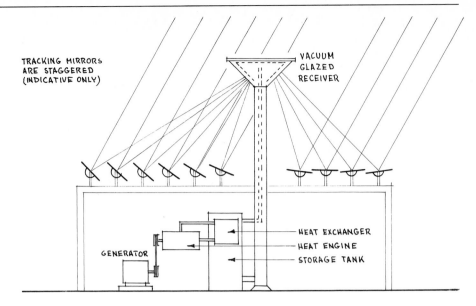

11.6
small-scale roof-mounted 'power-tower'
(Solar Physics Corporation)

quoted for small orders. This has since been reduced to about £9 (US $15·30) per watt.

More radical cost reductions have been achieved by two methods:

1 polycrystalline silicon cells, with increased (above 10%) efficiency. These are now available (from Solarex Corporation) for about £4 ($7) per watt.

2 edge-defined grown ribbon monocrystal cells (mentioned already in section 2.2). Originally developed at the Comsat laboratories, now the Japanese Toshiba Company has produced an 800 mm 'pull' of 30 mm width and 0·3 mm thickness, single crystal and is predicting a price reduction by a factor of 100.

The target of the ERDA program is to produce such cells at 50 cents ($\simeq$ 28p) per watt by 1985. A Bill was passed in the US in July 1976 authorising the construction of a 10 MW station using silicone photoelectric devices. This is expected to give a significant boost to the industry, which at present has an annual production rate of about 300 kW and the largest single order so far has been for cells of 180 kW. The annual production growth rate in recent years has been about 250% (or a factor of 2·5)—this order will induce a jump rather than growth.

Another development in photoelectric devices is the STEM (solar-thermal-electric module) developed at the University of Queensland. Essentially this is a medium concentration rate (up to 16) optical concentrating device, with a water cooled strip of silicone cells at the focal line. The output per cell area can be increased by a factor of 10, provided the temperature is kept around 60–70 °C. Hot water is generated as a by-product.

11.8
Solar houses

Shurcliff [8] lists 163 solar heated buildings, giving a brief description of each. 136 of these are in the USA. Only 19 of the 136 have been reviewed in part 5. Almost all of the remaining 117 houses have been built since 1974.

This survey is however far from complete. It has been suggested recently (January 77) that there are over 1000 solar houses in the US alone. A more comprehensive register is likely to be published in the near future by the Environmental Design and Research Centre (of Boston, Mass).

One of the most interesting and best documented projects is that carried out by the Colorado State University Solar Energy Applications Laboratory. Three identical houses have been built at Fort Collins (near Denver). One has been described in section 5.39. The second one has an air heater solar collector, made of galvanised steel, with non-selective black surfacing. The collector is double glazed. A fan driven by a two-speed 1·5 kW motor delivers air through the collectors at the rate of either 1 m³/s or 0·7 m³/s. Storage is provided by 16 tonnes of 50 mm diameter crushed rock. Auxiliary heating is supplied by a gas furnace. In summer operation outdoor air is circulated through the rock storage, after it has been evaporatively cooled. During the day room air is circulated through this cold store.

The third house has practically the same system as the first (Fig. 5.70), only the collector is different. Evacuated glass tube collectors have been used, supplied by Corning (cf 11.2 above). The tubes are 3 m long, 100 mm diameter, housing a finned tube absorber which has a selective black coating. These can supply temperatures up to 120°C. Tank insulation has been increased to 200 mm fibreglass.

A performance report [9] prepared after six months of operation of the first of the three buildings shows that 40% of the cooling, 86% of the heating and 68% of the domestic hot water was supplied from the solar source. The daily collection efficiency varied between 30 and 35%.

Most of the solar houses built in this period are different combinations, permutations and sizes of the houses and equipment described in part 5. They vary from *domestic hot water* installations (eg a 16-storey apartment block for the elderly at Brookline, Mass) through *heating* installations (eg the Norris Cotton Federal Building in Manchester, NH, with 33% solar contribution or the New York telephone exchange at Cutchogue, NY, giving 70% solar contribution), to buildings with *heating and air conditioning* installations (eg the NASA solar house at the Marshall Space Flight Centre, Alabama, with 50% summer and 100% winter solar contribution).

Many of these tackle the architectural problem only: how to incorporate and integrate the large collectors and storage tanks with the building design. Most involve serious system-optimisation exercises and produce refinements and technological perfection of the systems. Some are just 'bigger and better' variants of what has been done before.

Probably the most ambitious plan is the 56-storey Citicorp Centre in New York, which is to have some 1860 m² collector panels to power the heating, cooling and hot water systems.

There is, however, an increasing realisation that the thermally efficient design of the building can give a greater contribution (ie reduction in energy requirement) than an installed 'plumbing-type' solar heating system. The US Office of Emergency Preparedness [10] coined the phrases 'belt tightening' and 'leak plugging', denoting the possible ways of reducing energy usage. The former would mean a reduction of environmental standards (eg lower setting of thermostats, a lesser level of illuminance) whereas the latter would be the reduction of wastage (eg better insulation) or improvement of system efficiency. An 'active' solar energy system should only be relied on to supply the remaining need after all other energy saving measures have been made use of.

It has been demonstrated [11] that very simple modifications can reduce the heating and cooling requirements in an average suburban house in Melbourne to less than half. Whilst thermal insulation is important, thermal capacity provides greater benefits, where the diurnal variation in *sol-air temperature* is greater than about 20 deg C. Indeed a high thermal capacity building can be considered as a 'passive' solar heating system, akin to the houses of Curtis (5.11), Morgan (5.19), Michel & Trombe (5.22) or Hay (5.26).

Several houses have been built in Europe in the past two years, following such principles. [12]

The *zero energy house* at Lyngly, Denmark, has a south-facing vertical collector of 42 m² and attempts inter-seasonal heat storage in a 30 m³ steel cylindrical tank, located in the basement, having 600 mm mineral wool insulation. The tank temperature in summer exceeds 90 °C and is depleted to 45 °C in winter. The whole house has 300–400 mm mineral wool insulation and a heat recovery system is employed in its ventilation.

The *Termoroc house* at Malmö, Sweden, has a collector tilted at 72°, assisted by a reflector on the flat roof in front of it. This is expected to give a +15% contribution. It has 6 m³ storage tanks and an embedded coil floor- and ceiling-warming system.

The *Philips house* at Aachen, Germany, is a very complex and sophisticated system. The intention was to use evacuated glass tube collectors, but due to technological problems these have been replaced by a flat plate absorber and the vacuum tubes are only used as the transparent cover. It has a 42 m³ inter-seasonal heat storage tank. Various heat pumps are used to extract heat from waste water, from the earth below the house and from the main storage, after it has been depleted to below usable temperatures.

In all three houses the insulation is of an exceptionally high standard, reducing the heating requirement to less than one third of the normal.

11.9
Industrial applications

A survey carried out by CSIRO [13] showed that about $\frac{1}{3}$ of the total primary energy consumption in Australia is used to produce heat below 120° and a very large part of this is below 80 °C. This is particularly so in the food processing industry. Such low grade heat can be produced successfully and economically by solar heaters. A prototype installation is now in progress at a soft drink manufacturing plant near Canberra. This uses 77 m² double glazed collectors (10 rows of 5 panels, each 1210 × 1365 mm) and a storage tank of 21 m³ capacity.

A recent study by the Western Australia State Energy Commission [14] examined the use of solar energy in the mineral processing industries. It concluded that solar energy could be used for the following processes:

—to produce hot water for leaching
—to produce low grade steam
—to evaporate water, producing saturated solutions
—to dry various metal concentrates before smelting or in preparation for roasting or sintering
—to pre-heat boiler water, or to produce distilled boiler water in areas where the available water is unsuitable.

These would be applicable to nickel, copper, lead, silver and zinc processing as well as in the production of alumina and even in uranium processing!

11.10
Fuel production

Fuels possess energy in chemical form, which can be released when and where it is required, thus the production of fuel materials would provide the most convenient solution to the energy storage problem.

Hydrogen can be produced by electrolysis of water. The electricity for this purpose can be produced by solar energy. There is, however, a more direct way of producing hydrogen, pursued by the Australian National University [15]: ammonia dissociation under high pressure by direct thermal means, at the focal point of parabolic collectors, at about 700°C. The resulting hydrogen and nitrogen can be stored separately or piped to the point of use, and made to re-combine in the presence of a catalyst. This is a highly exothermic process, producing 450–550°C temperatures.

The term 'biomass conversion' is often used to describe the growing of plants (using solar energy through photosynthesis) and subsequent processing of the plant material to produce a convenient fuel. The efficiency of the photosynthentic process (cf section 2.1) seldom exceeds 1 or 2%. Normal agricultural efficiencies in Europe or North America are less than 0·4%. The highest recorded value has been obtained in northern Australia with bullrush millet: 4·2% over a short period. Research is being conducted at several centres into increasing plant efficiency.

A survey of plausible plants and conversion methods was carried out recently at the Energy Research Centre of Sydney University. [16] In northern Australia tropical crops, such as cassava (tapioca, manioc), kenaf (an annual fibre crop), elephant grass and sugar cane are the most promising. The first one gives starch in underground tubers and cellulose in its leafy tops. Kenaf and elephant grass yield mainly cellulose, whilst sugar cane gives both sucrose and cellulose. In southern, temperate Australia, forest production of some eucalyptus species would give better results.

Three major types of conversion processes are feasible:

1 hydrolysation of starch (an inexpensive process) or of cellulose (more difficult and expensive) and subsequent fermentation of the sugars produced, to give ethyl alcohol. The energy efficiency of this process is at the very best 17%, and with some celluloses it can be negative. (This is the ratio of the calorific value of the fuel produced to the sum of the calorific value of raw plant material + any external energy input).

2 bacterial fermentation of plant material to produce bio-gas (mainly methane). The energy efficiency of this process can reach 34%, but the carbon/nitrogen ratio of the raw material is critical

3 pyrolysis to produce oil: the photosynthetic material is shredded, dried, pulverised and flash pyrolysed at 500°C (ie heated in the absence of oxygen). The energy efficiency can be as much as 52%.

None of these methods are economically competitive with conventional fuels at present prices, except if a no-cost byproduct is used (eg cereal straw) as raw material, where it is available, ie if there is no transportation cost.

The CSIRO Solar Energy Studies Unit has also examined solar ethanol (ethyl alcohol) production. Their conclusions are somewhat more optimistic than the above, at least in relation to the hydrolysis of wood. It is suggested [17] that the process would yield 3 units of energy in liquid fuel for 1 unit of energy input. About 10 tonnes of wood are needed to produce 1 tonne of ethanol. The overall energy efficiency would be about 20%. The cost of ethanol thus produced would be about double the cost of coal-based liquid fuel or three times the present price of petrol.

References

1 Scoville, A E
An alternative cover material for solar collectors
paper 30/11 in ISES Congress, Los Angeles, 1975

2 Cathro, K J and Christie, E A
The application of chrome black coatings to solar panels
in ISES-ANZ Section Symposium, Melborne, November 1976

3 Szulmayer, W
A solar strip concentrator
Solar Energy 14 (327) 1973

4 Szulmayer, W
Stationary solar concentrators for industrial heating and cooling
in Heliotechnique and Development conf. Dhahran, Saudi Arabia
November 1975 (Development Analysis Associates, Cambridge, Mass)

5 Faber, E A et al.
Solar operation of ammonia/water air conditioners
paper 44/7 in ISES Congress, Los Angeles, 1975

6 Barber, R E
Design and test of a prototype 3-ton solar heated Rankine-cycle air conditioner
paper 44/10 in ISES Congress, Los Angeles, 1975

7 see photo in *National Geographic Magazine 149* (380) March 1976

8 Shurcliff, W A
Solar heated buildings: a brief survey
10th edition, September 1975 (13th 'final edition' was published in January 1977)

9 Ward, D S and Löf, G O G
Design, construction and testing of a residential solar heating and cooling system
progress report to NSF and ERDA
1 July 74–1 February 75
Colorado State University, June 1975

10 Office of Emergency Preparedness
The potential for energy conservation
Washington DC, 1972

11 Williamson T and Coldicutt, A B
Comparison of performances of conventional and solar houses
symposium on Solar Energy Utilisation in Dwellings
Inst. Mech. Engr. Melbourne. November 1974

12 see eg *Architects' Journal*, April 21, 1976, p. 780

13 Morse, R N
Solar energy as a major source of power for Australia
Inst. of Engr. conference, Australia, 1974

14 Saunders, D W
Industrial and mineral applications of solar energy
in ISES-ANZ Section Symposium, Melbourne, November 1976

15 Carden, P O et al.
Thermochemical energy transfer and storage
in ISES-ANZ Section Symposium, Canberra, November 1975

16 McCann, D J and Saddler, H D W
Photobiological energy conversion
in ISES-ANZ Section Symposium, Canberra, November 1975

17 Siemon, J R
The production of solar ethanol from Australian forests
CSIRO Solar Energy Studies Report No. 75/5

Part 12   Theory and Methods

12.1
Thermal processes in the collector

A reasonable consensus seems to have emerged regarding the description of thermal processes in a collector. [1] This is only slightly different from the introduction given in sections 2.3 and 2.4.

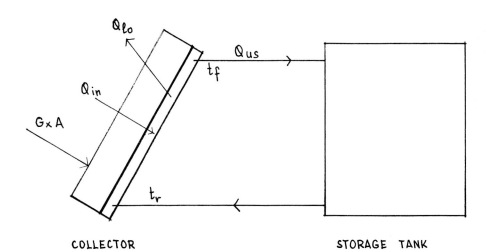

12.1
thermal quantities in collection

With reference to Fig. 12.1 we can say that with a total radiation intensity on the collector surface of G W/m² (in our earlier notation: I_{pt}), if the collector area is A m², it will receive energy at its outer surface at the rate of G × A watts. The heat input rate into the actual absorber surface will be

$$Q_{in} = F' (\theta \times a)e \times G \times A \qquad \qquad ...1)$$

where θ = transmission coefficient of glass cover
      a = absorption coefficient of plate
      F' = plate efficiency (transfer factor from plate surface to fluid)
      e = 'effectiveness' to allow for dirt on the glass, shading by edges or glazing bars, but also for re-radiation effects by the heated glass towards the absorber
      $(\theta a)e$ is often referred to as the *effective transmittance-absorbance product* often denoted as $(\tau\alpha)_e$.

There will be a heat loss rate from the collector

$$Q_{lo} = A \times U(\bar{t}_w - t_{eq}) \qquad \ldots 2)$$

where U = overall heat loss coefficient, fluid to environment

$\bar{t}_w$ = mean water temperature $= \dfrac{t_r + t_f}{2}$

t_r = temperature of return to collector

t_f = temperature of flow from collector

t_{eq} = equivalent surround temperature taken as

$t_o - 3$ to allow for radiant losses to the sky

t_o = outdoor air temperature.

Thus the useful heat transported away from the collector is

$$Q_{us} = Q_{in} - Q_{lo} \qquad \ldots 3)$$

The instantaneous efficiency of collection is defined as 'useful heat over radiation received'

$$\eta = \frac{Q_{us}}{G \times A} \qquad \ldots 4)$$

Substituting from equations 3, 1 and 2

$$\eta = \frac{F'\,(\theta a)e \times G \times A - A \times U\,(\bar{t}_w - t_{eq})}{G \times A}$$

which, by cancellations becomes

$$\eta = F'\,(\theta a)e - \frac{U\,(\bar{t}_w - t_{eq})}{G} \qquad \ldots 5)$$

When there is no loss, thus $Q_{us} = Q_{in}$, which occurs when $\bar{t}_w = t_{eq}$, we get the fundamental efficiency (or 'no-loss' efficiency)

$$\eta_0 = \frac{Q_{in}}{G \times A} \qquad \ldots 6)$$

Substituting from equation 1

$$\eta_0 = \frac{F'\,(\theta a)e \times G \times A}{G \times A}$$

$$\eta_0 = F'\,(\theta a)e \qquad \ldots 7)$$

Finally, substituting into equation 5

$$\eta = \eta_0 - \frac{U\,(\bar{t}_w - t_{eq})}{G} \qquad \ldots 8)$$

It has been observed that for a given wind velocity the value of U is not constant, but changes as a function of the temperature difference, as

$$U = a + b\,(\bar{t}_w - t_{eq}) \qquad \ldots 9)$$

where a and b are constants, depending on physical properties of the particular collector.

12.2
Collector testing

With the proliferation of solar heaters produced by a large number of different manufacturers, the need has arisen for a standard method of testing and describing the performance of a particular collector. Single figure efficiency factors are useless, as they vary with incident radiation, inlet-outlet temperatures and ambient temperature.

Israel has had standards both for water heater construction and quality (No. 579) and for testing methods (No. 609) since 1966.

The National Bureau of Standards (USA) has produced an interim report (NBSIR 74-635), ie a proposed standard. According to this, the efficiency (η) is expressed as a function of $\dfrac{\Delta t}{G}$ for a range of values of the latter where $\Delta t = \bar{t}_w - t_o$.

The performance of the collector is tested at various pre-determined inlet (return) water temperatures, these η values are plotted and a *collector characteristic curve* is proposed.

The Standards Association of Australia has a subcommittee currently working on a standard proposal, which is likely to be a slightly modified version of the American method. In fact, the only major modification is that here the temperature difference will be taken as

$$\Delta t = \bar{t}_w - t_{eq} \quad (t_{eq} \text{ being } t_o - 3).$$

The standard test method uses a rig such as that shown diagrammatically in Fig. 12.2.

Tank 1 provides a constant head.
Tank 2 provides water at a set temperature, being continuously mixed with water in tank 1.

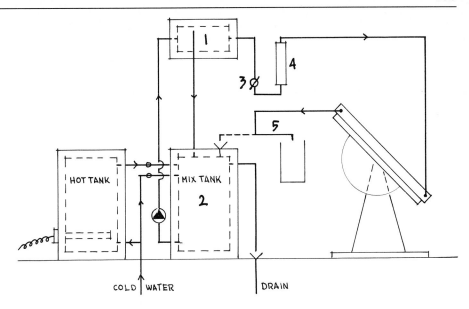

12.2
collector test-rig

Flow through the collector to be tested is regulated by needle valve 3 and measured by flow meter 4, but fine measurement during the 10-minute actual testing is done by batching through swivel spout 5. Efficiency is measured at a minimum of four different values of inlet water temperature, preferably being 10, 30, 50 and 70 degC above equivalent surround temperature. Measurement at each of these temperatures is repeated four times, to obtain four *data points*.

After the temperature has been set, flow through the collector is to be allowed for 30 minutes (at each of the temperature levels to be used). Each data point will be based on 10 minute integrated values. The fluid flow rate is to be approximately 0·01 kg/m²s with a water collector (0·01 m³/m²s with an air heater).

During the 10-minute test the average flow temperature is obtained. This is subtracted from the known return (inlet) temperature, to get the increment. This is multiplied by the measured mass flow and the specific heat capacity of water to get Q_{us}. This is divided by the product of the measured G and A (area) to get η (as equation 4). Δt is established as

$$\frac{t_r + t_f}{2} - t_{eq}$$

and divided by the measured G (averaged for the 10 minutes). This value is located on the X axis and the value of η is plotted as the ordinate (Fig. 12.3).

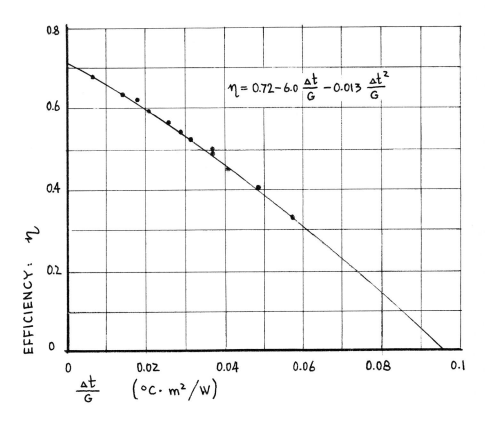

$$\eta = 0.72 - 6.0\,\frac{\Delta t}{G} - 0.013\,\frac{\Delta t^2}{G}$$

EFFICIENCY: η

$\frac{\Delta t}{G}$ (°C . m²/W)

12.3
collector characteristic curve

Where the curve thus produced intersects the Y axis, the value of η_0 is obtained.

A second order function can be fitted to the curve using the method of least squares, which will take the form [2]:

$$y = A - Bx - Cx^2$$

where $x = \dfrac{\Delta t}{G}$

and $y = \eta$

thus

$$\eta = A - B\frac{\Delta t}{G} - C\left(\frac{\Delta t}{G}\right)^2$$

putting η_0 for A, a for B and bG for C

we get the *collector characteristic function*

$$\eta = \eta_0 - a\frac{\Delta t}{G} - b\frac{\Delta t^2}{G} \qquad\qquad \ldots 10)$$

where the values of η_0 and the constants a and b constitute the test results.

(The last term would be $bG\left(\dfrac{\Delta t}{G}\right)^2$ which simplifies to $b\dfrac{\Delta t^2}{G}$ where b can be found as $b = \dfrac{C}{G}$ using G averaged over all tests.)

Combining equations 8 and 10

$$\frac{U\,(\bar{t}_w - t_{eq})}{G} = a\frac{\Delta t}{G} + b\frac{\Delta t^2}{G} \qquad \text{as } \Delta t = \bar{t}_w - t_{eq}$$

$$U = a + b\,(\bar{t}_w - t_{eq})$$

which is the same expression as equation 9. Thus from the test results both parameters η_0 and U needed for performance prediction can be established.

Some typical test results are

a) $\eta = 0\cdot61 - 2.18\dfrac{\Delta t}{G} - 0\cdot015\dfrac{\Delta t^2}{G}$

b) $\eta = 0\cdot69 - 3\cdot30\dfrac{\Delta t}{G} - 0\cdot018\dfrac{\Delta t^2}{G}$

c) $\eta = 0\cdot71 - 5\cdot15\dfrac{\Delta t}{G} - 0\cdot017\dfrac{\Delta t^2}{G}$

d) $\eta = 0\cdot72 - 8\cdot00\dfrac{\Delta t}{G} - 0\cdot018\dfrac{\Delta t^2}{G}$

12.3 Simulation

In the absence of such test results, for the purposes of performance simulation in the crudest form, a single figure efficiency factor has been assumed. In a slightly more sophisticated method the value of η_0 was calculated (from equation 7 above) by using published values of F', θ, a and e and a U-value was also taken from available literature. A great deal of sophistication can be introduced to compute the values of all these constants [3] but the results are not very reliable, partly because of differences between a real collector and its geometrical abstraction, partly due to the fact that certain basic constants always have to be assumed. Performance simulation on the basis of a collector characteristic curve is far more reliable.

The most sophisticated computer simulation system has been developed at the University of Wisconsin, known as TRNSYS [4] (transient system simulation), written in FORTRAN. It is actually an organizational system, an 'executive program', which allows a large variety of modules or subroutines to be plugged in corresponding to the physical system to be analysed. The analytical depth of the program is such that its operation is prohibitively expensive in terms of computer time.

Many other researchers have developed programs of their own.* At the other end of the scale is a simplified program developed at the Architectural Science Unit of the University of Queensland for use with a programmable calculator (Canola SX300, having 500 memories and taking up to 5000 program steps). No programming language is necessary. It is a modular program, where each module is stored on magnetic tape. When inputting the selected modules, one extra instruction is sufficient to provide continuity. So far about a dozen modules are available, such as

—solar position angles
—incident radiation on tilted planes
—collection, pumped, into stratified tank
—collection, thermosyphon, into stratified tank
—effect of load on storage temperatures
—auxiliary heater in tank

*Eg 1) Los Alamos Scientific Laboratory (New Mexico): continuous dynamic simulation
2) Sandia Laboratories, Albuquerque (New Mexico): SOLSYS program, an executive program+a library of subroutines
3) Colorado State University, Solar Energy Applications Laboratory: a simple stochastic modelling system (ie using a probability-based error elimination process)
4) University of Melbourne, Dept. of Mech. Engr.: modelling of the Michell-Trombe passive system, based on a finite difference method.

—space heating requirement
—auxiliary heater in series (space heating)

but further ones are being developed.

12.4
Radiation data

Any such simulation is meaningful only if it is based on measuring solar radiation data. Such data is normally available only for a horizontal plane and in most cases only the total radiation is measured. Radiation on a tilted plane can be calculated only if the direct and diffuse components are known, using the method described in section 1.9.

The CSIRO has developed an empirical method for the estimation of the two components [5] if the measured total is known. The clear sky solar radiation on the horizontal surface is calculated on the basis of the solar constant, sun position angles and atmospheric precipitable water content, using a method similar to Spencer's. [6] This calculated value (T) is compared with the measured horizontal total (G).

The ratio G/T is indicative of sky conditions, thus the proportion of total that will be diffuse (D) can be deduced

if $0 < \dfrac{G}{T} \leqslant 0.4$ heavy cloud exists $D = 0.94\, G$

if $0.4 < \dfrac{G}{T} \leqslant 1$ partly clouded $D = \dfrac{1.29 - 1.19\, G/T}{1 - 0.334\, G/T}\, G$

if $\dfrac{G}{T} > 1$ highly reflective clouds exist $D = 0.15\, G$

A computer program PRERAD has been produced to do this calculation and continue with calculating values of radiation incident on any tilted plane.

12.5
Economic analyses

The Australian Consumers' Association journal CHOICE published an assessment 'Solar heating for hot water' in August 1975, comparing this with the cost of electric water heating. Their argument was built up as follows:

1   The cost of electricity in Australia varies from 0.99 cents to 3.15 cents per kWh.

2   If the daily hot water consumption is 91 litres and this is to be heated from 15° to 70°C, in an all-electric system the annual cost will vary from $31 to $102.

3   Half of this would be supplied by the solar collector to a value of $15.50 to $51.00.

4   The price of an average solar hot water system may be $500.

5   If such a sum were to be invested at 12% interest, it would earn $63.40 per year (compounded monthly) — which is more than what the solar heater would save, even in high tariff areas.

There are many errors and fallacies in this argument.

Re **2**   The daily hot water consumption of a 4-person family is at least double the amount shown, thus the annual cost will vary from $62 to $204.

Re **3**   Extensive in situ studies showed that the solar contribution is not 50%, but varies between 61% and 81% and in many areas the auxiliary heater is not used at all (ie 100% solar contribution). Taking 70% as a conservative mean value, the annual saving would be $43.40 to $142.80.

Re **4**   In such a comparison not the full price of the solar unit, but the difference between a solar and a conventional unit should be used, say $400.

Re **5**   Thus the interest earned would be $50.75 only. As income, this would be taxed at 42.5%, thus the net earning would be $29.18 — which is less than what the solar heater would save, even in the lowest tariff areas.

This is a rather simplistic analysis, which—based on the right facts—can be quite convincing, but using the wrong assumptions (due to either ignorance or hostility) it can do a great harm. Although the latter figures give a positive result, this method still does not take into account inflation, which would give a further advantage to the solar water heater.

In section 7.7 a more sophisticated method of economic analysis has been introduced, using a discounting method and the concept of *present worth*. It has also been suggested that in an inflationary situation only the difference between inflation rate and interest rate should be used in the formula given. This is an oversimplification. It may be worth while reviewing the principles of this calculation, as described in reference [7].

The basic compound interest formula is
$A = P\,(1 + i)^y$ . . . a)

where A = amount in future (capital + compound interest)
P = present amount (capital)
i = annual interest rate (as decimal fraction)
y = number of years.

Conversely the present worth of amount A paid y years hence:

$P = A \dfrac{1}{(1+i)^y}$ which can be written as

$P = A(1+i)^{-y}$ ... b)

The present worth of a sum A payable annually over y years is

$P = A \dfrac{(1+i)^y - 1}{i(1+i)^y}$ which is the expression given on p. 125

this can also be written as

$P = A \dfrac{1 - (1+i)^{-y}}{i}$ ... c)

If an inflation rate of f is allowed for, the above expressions will become

$A = P \left(\dfrac{1+i}{1+f}\right)^y$ ... a¹)

$P = A \left(\dfrac{1+i}{1+f}\right)^{-y} = A \left(\dfrac{1+f}{1+i}\right)^y$ ... b¹)

$P = A \dfrac{1 - \left(\dfrac{1+f}{1+i}\right)^y}{i-f} = A \dfrac{1}{i-f} \left(1 - \left(\dfrac{1+f}{1+i}\right)^y\right)$ ... c¹)

However in an inflationary situation the annual amount A is not constant either, it has to be multiplied by (1+f) thus the final expression becomes

$P = A \dfrac{1+f}{i-f} \left(1 - \left(\dfrac{1+f}{1+i}\right)^y\right)$ ... d)

Using this expression the term A should mean the annual cost in terms of present day prices.

12.6 Feasibility criteria

In section 7.7 the value of annual *auxiliary heating cost* was discounted and added to the capital cost—for various collector areas. The minimum point of the resultant curve was found to define the optimum collector area.

The same discounting method can be used to find the present worth of the *annual saving in heating cost* due to the solar installation (ie the value of the solar contribution). If the capital cost is deducted from this present worth, the *net present value* (NPV) is obtained. From equation (d) this statement can be written as

$NPV = -C + a \dfrac{1+f}{i-f} \left(1 - \left(\dfrac{1+f}{1+i}\right)^y\right)$

where C = capital cost (or extra capital cost).

We can substitute F for the right hand side expression

$F = \dfrac{1+f}{i-f} \left(1 - \left(\dfrac{1+f}{1+i}\right)^y\right)$ (if i = f then F = y)

The criterion of feasibility is that the NPV should have a positive value which can be written as

$A \times F > C$ or

$\dfrac{C}{A} < F$

This can be linked to the figure of merit concept introduced in section 7.1, which has been defined as

$FM = \dfrac{A \times 10}{C}$ where 10 was an arbitrary number of years.

Substituting F for the above 10

$FM = \dfrac{A \times F}{C}$

Thus the figure of merit becomes a definite criterion of feasibility:

$FM > 1$.

12.7 Heat tables

A paper of the CSIRO Solar Energy Studies Unit [8] describes a method of predicting the heat production of a particular collector in a specific location, and presenting it in a form convenient for use by practising 'solar engineers'.

The collector characteristic curve (η versus $\Delta t/G$) must be known, as well as detailed climatic data of the location (solar radiation, temperature and wind velocity—hourly values). The collector tilt and orientation and the water flow rate must be assumed.

The computer program SOCOFI will calculate the daily total heat production at various outlet temperatures and will produce the average daily heat production (in MJ/m²) for every month over a number of years (depending on the availability of climatic data—normally 3–10 years in Australia). For this average day of each month it will print out the highest value over the years, the mean value and the lowest value. These printouts are referred to as *heat tables*.

The design engineer will have calculated his water heating requirement and established his operating temperature. He can look up the expected collection values and divide it into the heat requirement to get the required collector area. It must be his value judgement whether to install collectors enough to supply all the heat in the worst month (ie a high percentage of solar contribution, but a low utilisation rate) or to install collectors of a lesser area, giving a lower percentage of contribution, but a better utilisation rate. For such a decision he may use the economic analysis methods described in the previous sections, relying on the heat tables to predict the amount of energy contributed, thus the value of the benefit of various collector areas.

12.8 Influence of use-pattern

Any economic analysis is based on the value of solar contribution, which in turn depends on the prediction of the performance of the proposed solar device.

Recently two members of the same Energy Advisory Council* produced studies on the economic feasibility of solar water heaters and reached diametrically opposite conclusions. The main reason for this is that both used an annual average efficiency figure, and obviously different figures. Both are equally unreliable.

Firstly, the value of solar contribution depends on the amount of hot water consumed. Secondly, the efficiency of collection depends on the collector inlet (return) water temperature, thus on the state of water at the bottom of the tank. Thirdly, the amount of electricity used by the off-peak electric booster element depends on the tank temperature at the time when the time switch is on. Thus the expected performance very strongly depends on the hourly hot water use-pattern.

If most of the hot water is used in the morning (eg, if all members of the family have a shower after getting up and any washing is done in the morning) the tank will contain mostly cold water when the solar collection begins. The collection efficiency will thus be high and by the evening there will be a tankful of hot water. If only little consumption occurs in the evening, the thermostat of the electric booster may not switch on at all.

On the other hand, if most of the hot water is used in the evening (eg, if the wife works, thus the washing is done after she comes home, if all members of the family have baths or showers in the evening) the resulting tankful of cold water will be heated up, after sunset, by the electric booster. As there is little consumption of hot water in the morning, the collection starts with a tankful of hot water, thus its efficiency will be low.

A reported simulation study [9] was based on the following assumptions:

225 litre daily consumption: thermostat setting—55°C: booster—1800 W, located $\frac{1}{3}$ of the tank height; off-peak time—20·00 h to 6·00 h: location—Brisbane: date—June 24.

Use pattern I: 168 litre used before 10·30 h, the remainder spread between midday and evening.

Use pattern II: 35 litre used in early hours, 90 litre around midday, 100 litre after 18.00 h.

Use pattern III: 35 litre used in early hours, 20 litre at midday, 170 litre after 17.00 h.

The following results were obtained:

use pattern	electricity used	net* solar collection
I	0·354 kWh	9·718 kWh
II	3·470	9·540
III	5·184	8·190

*after deducting tank heat losses

The gross solar collection in pattern I is much greater, but the evening tank temperature is far above the thermostat setting, thus the heat losses reduce the net figure. The most significant difference is in electricity use:

a factor of 10 between I and II and
a factor of 14 between I and III.

*of an Australian state government

References

1 Duffie, J A and Beckman, W A
Solar energy thermal processes
J Wiley (New York) 1974

2 Pott, P and Cooper, P I
An experimental facility to test flat plate solar collectors
CSIRO, Div. Mech, Eng. technical report TR 9 (1976)

3 Jordan, R G (ed)
Low temperature engineering application of solar energy
chapter III: Design factors influencing collector design
by A Whillier
ASHRAE, 1967

4 Klein, S A et al.
A method of simulation of polar processes and its application
Solar Energy 17 (29) 1975

5 Bugler, J W
The determination of hourly solar radiation incident upon an inclined plane from hourly measured global horizontal insolation
CSIRO, Solar Energy Studies, report 75/4 (1975)

6 Spencer, J W
Estimation of solar radiation . . . on clear days
CSIRO, Div. Bldg. Res., paper 15 (1965)

7 Hawkins, C J and Pearce, D W
Capital investment appraisal
Macmillan (London) 1971

8 Salt, H
A note on solar collector performance for engineering use
CSIRO, Solar Energy Studies, 1/7/1975

9 Szokolay, S V
Modelling of solar collection
ANZ Architectural Science Assoc. conf. 1976 Canberra

Appendix 1

Optimisation of collector tilt

The program is based on hourly data of total and diffuse radiation intensity, measured on a horizontal plane. Loudon's 'augmented direct+background diffuse radiation' technique*is used. On the basis of numerous studies by others as well as his own measurements he has established that the intensity of 'background diffuse radiation' is a function of solar altitude only, thus it does not depend on sky conditions. The remainder of the measured diffuse intensity is taken as 'circumsolar diffuse radiation' and is added to the direct component, to be handled vectorially, as 'augmented direct radiation'.

The program calculates the intensities (direct and diffuse) incident on the particular tilted plane, applies the appropriate transmission and absorption coefficients (both varying with the angle of incidence) thus it produces the quantity absorbed by the plate.

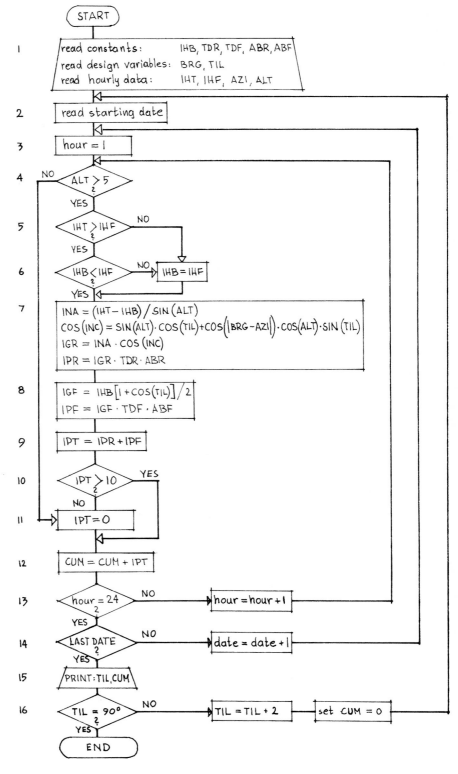

COMPUTER FLOW DIAGRAM : OPTIMISATION OF TILT

* A G Loudon: The interpretation of solar radiation measurements for building problems. In 'Sunlight in Buildings', CIE Newcastle Conference.

The cumulative total absorbed amount of radiation is calculated for the year, or any other defined period. After this the tilt is increased by 2° and the process is repeated.

No radiation is taken into account when the solar altitude angle is less than 5° (step 4) or when the absorbed value would be less than 10 W/m².

Symbols used

Constants: IHB   intensity, horizontal, background diffuse (W/m²)*

ALT:	0°	5°	10°	15°	20°	25°	30°	35°	40°	45°	50°	55°	60°
IHB:	0	28	43	54	63	71	77	83	88	93	99	101	105

TDR   transmittance for direct radiation (double glazing)

INC:	0°	10°	20°	30°	40°	50°	60°	70°	80°	90°
TDR:	0·72	0·72	0·72	0·71	0·70	0·67	0·60	0·48	0·28	0

TDF   transmittance for diffuse radiation = 0·60

ABR   absorption coefficient for direct radiation

INC:	0°	10°	20°	30°	40°	50°	60°	70°	80″	90°
ABR:	0·98	0·97	0·95	0·94	0·93	0·90	0·85	0·78	0·55	0

ABF   absorption coefficient for diffuse radiation = 0·95

Design variables:
BRG   bearing, orientation of plate (re N = 0°)

TIL   tilt of plate from horizontal
(in this program initially set at the lowest angle to be considered)

Hourly data:
IHT   intensity, horizontal, total (W/m² or Wh/h.m²)

IHF   intensity, horizontal, diffuse (W/m² or Wh/h.m²)

AZI   solar azimuth angle (re N = 0°) ⎫
ALT   solar altitude angle ⎬ at hour—30 minutes

Other symbols:
INA   intensity, normal, augmented direct (W/m²)

IGR   intensity on glass cover, direct

IGF   intensity on glass cover, diffuse

IPR   intensity absorbed by plate, direct

IPF   intensity absorbed by plate, diffuse

IPT   intensity absorbed by plate, total

CUM   cumulative value of IPT (Wh/m²)

* Loudon's values for cloudless sky have been adopted, as his values for partially clouded sky are virtually never exceeded by the IHF values of meteorological records. If IHB is greater than IHF, the 'circumsolar diffuse' would have a negative value, hence the check in step 6. When IHT = IHF in the meteorological data, there is no directional component (see step 5).

Appendix 2

System simulation

The program simulates the behaviour of the system shown below (which is an earlier version of that shown in Fig. 5.55). Its base data are the hourly IPT values produced by a slightly modified version of the above program and hourly temperature values.

However, whilst the optimisation program can be based on hourly data averaged for 10-day periods and over several years, in this program hourly simultaneous radiation and temperature values for every day of an actual year are used.

Settings of thermostats and controls are put in as 'design variables', together with the collector area, the specific heat loss rate of the building, the volume of storage and an assumed hot water consumption pattern.

The net heat gain of the collector is calculated, allowing for heat losses from the collector. This amount enters the storage. The space and water heating demand is established. Flow and return temperatures are calculated. The central part of the program is the simulation of stratification in the storage tank (devised by David Hodges).

Explanation of the symbols used follows the flow diagram.

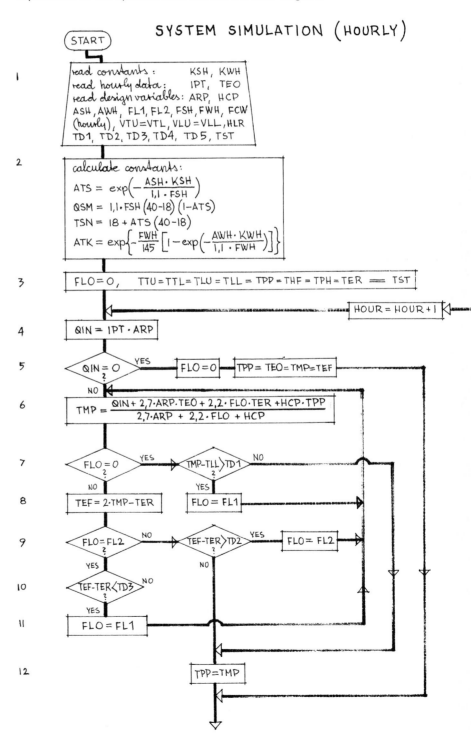

SYSTEM SIMULATION (HOURLY)

START

1. read constants: KSH, KWH
read hourly data: IPT, TEO
read design variables: ARP, HCP
ASH, AWH, FL1, FL2, FSH, FWH, FCW (hourly), VTU=VTL, VLU=VLL, HLR
TD1, TD2, TD3, TD4, TD5, TST

2. calculate constants:
$$ATS = \exp\left(-\frac{ASH \cdot KSH}{1.1 \cdot FSH}\right)$$
$$QSM = 1.1 \cdot FSH (40-18)(1-ATS)$$
$$TSN = 18 + ATS (40-18)$$
$$ATK = \exp\left\{-\frac{FWH}{145}\left[1-\exp\left(-\frac{AWH \cdot KWH}{1.1 \cdot FWH}\right)\right]\right\}$$

3. FLO=0, TTU=TTL=TLU=TLL=TPP=THF=TPH=TER == TST

HOUR = HOUR + 1

4. QIN = IPT · ARP

5. QIN = 0 ? — YES → FLO=0 — TPP=TEO=TMP=TEF
NO

6. $$TMP = \frac{QIN + 2.7 \cdot ARP \cdot TEO + 2.2 \cdot FLO \cdot TER + HCP \cdot TPP}{2.7 \cdot ARP + 2.2 \cdot FLO + HCP}$$

7. FLO=0 ? — YES → TMP-TLL > TD1 ? — NO
NO / YES

8. TEF = 2·TMP−TER / FLO = FL1

9. FLO=FL2 ? — NO → TEF−TER > TD2 ? — YES → FLO=FL2
YES / NO

10. TEF−TER < TD3 ? — NO
YES

11. FLO=FL1

12. TPP=TMP

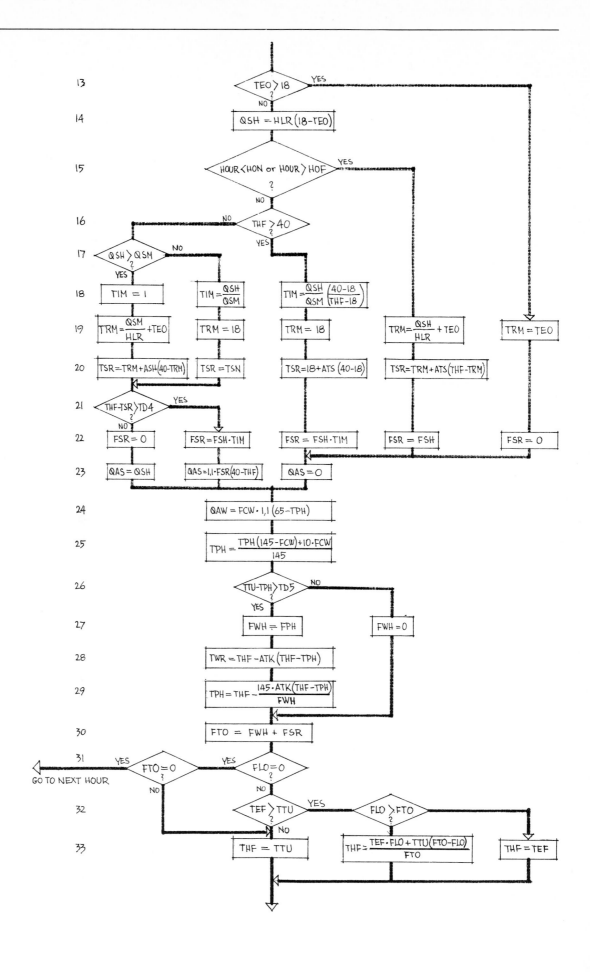

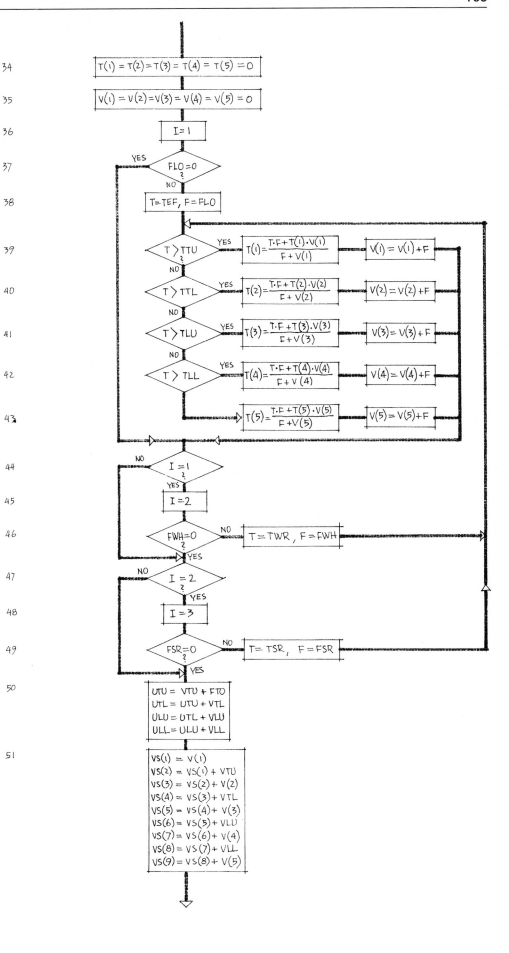

34 $T(1) = T(2) = T(3) = T(4) = T(5) = 0$

35 $V(1) = V(2) = V(3) = V(4) = V(5) = 0$

36 $I = 1$

37 FLO = 0 ? — YES / NO

38 $T = TEF, \ F = FLO$

39 $T > TTU$? — YES $T(1) = \dfrac{T \cdot F + T(1) \cdot V(1)}{F + V(1)}$ $V(1) = V(1) + F$ NO

40 $T > TTL$? — YES $T(2) = \dfrac{T \cdot F + T(2) \cdot V(2)}{F + V(2)}$ $V(2) = V(2) + F$ NO

41 $T > TLU$? — YES $T(3) = \dfrac{T \cdot F + T(3) \cdot V(3)}{F + V(3)}$ $V(3) = V(3) + F$ NO

42 $T > TLL$? — YES $T(4) = \dfrac{T \cdot F + T(4) \cdot V(4)}{F + V(4)}$ $V(4) = V(4) + F$

43 $T(5) = \dfrac{T \cdot F + T(5) \cdot V(5)}{F + V(5)}$ $V(5) = V(5) + F$

44 $I = 1$? — NO / YES

45 $I = 2$

46 FWH = 0 ? — NO $T = TWR, \ F = FWH$ / YES

47 $I = 2$? — NO / YES

48 $I = 3$

49 FSR = 0 ? — NO $T = TSR, \ F = FSR$ / YES

50 $UTU = VTU + FTO$
    $UTL = UTU + VTL$
    $ULU = UTL + VLU$
    $ULL = ULU + VLL$

51 $VS(1) = V(1)$
    $VS(2) = VS(1) + VTU$
    $VS(3) = VS(2) + V(2)$
    $VS(4) = VS(3) + VTL$
    $VS(5) = VS(4) + V(3)$
    $VS(6) = VS(5) + VLU$
    $VS(7) = VS(6) + V(4)$
    $VS(8) = VS(7) + VLL$
    $VS(9) = VS(8) + V(5)$

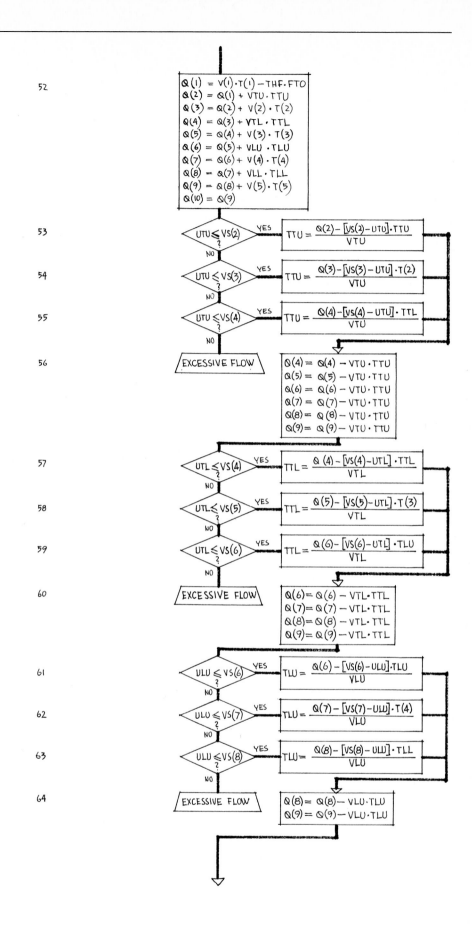

52

$$Q(1) = V(1) \cdot T(1) - THF \cdot FTO$$
$$Q(2) = Q(1) + VTU \cdot TTU$$
$$Q(3) = Q(2) + V(2) \cdot T(2)$$
$$Q(4) = Q(3) + VTL \cdot TTL$$
$$Q(5) = Q(4) + V(3) \cdot T(3)$$
$$Q(6) = Q(5) + VLU \cdot TLU$$
$$Q(7) = Q(6) + V(4) \cdot T(4)$$
$$Q(8) = Q(7) + VLL \cdot TLL$$
$$Q(9) = Q(8) + V(5) \cdot T(5)$$
$$Q(10) = Q(9)$$

53    $UTU \leq VS(2)$?   YES $TTU = \dfrac{Q(2) - [VS(2) - UTU] \cdot TTU}{VTU}$

NO

54    $UTU \leq VS(3)$?   YES $TTU = \dfrac{Q(3) - [VS(3) - UTU] \cdot T(2)}{VTU}$

NO

55    $UTU \leq VS(4)$?   YES $TTU = \dfrac{Q(4) - [VS(4) - UTU] \cdot TTL}{VTU}$

NO

56   EXCESSIVE FLOW

$$Q(4) = Q(4) - VTU \cdot TTU$$
$$Q(5) = Q(5) - VTU \cdot TTU$$
$$Q(6) = Q(6) - VTU \cdot TTU$$
$$Q(7) = Q(7) - VTU \cdot TTU$$
$$Q(8) = Q(8) - VTU \cdot TTU$$
$$Q(9) = Q(9) - VTU \cdot TTU$$

57    $UTL \leq VS(4)$?   YES $TTL = \dfrac{Q(4) - [VS(4) - UTL] \cdot TTL}{VTL}$

NO

58    $UTL \leq VS(5)$?   YES $TTL = \dfrac{Q(5) - [VS(5) - UTL] \cdot T(3)}{VTL}$

NO

59    $UTL \leq VS(6)$?   YES $TTL = \dfrac{Q(6) - [VS(6) - UTL] \cdot TLU}{VTL}$

NO

60   EXCESSIVE FLOW

$$Q(6) = Q(6) - VTL \cdot TTL$$
$$Q(7) = Q(7) - VTL \cdot TTL$$
$$Q(8) = Q(8) - VTL \cdot TTL$$
$$Q(9) = Q(9) - VTL \cdot TTL$$

61    $ULU \leq VS(6)$?   YES $TLU = \dfrac{Q(6) - [VS(6) - ULU] \cdot TLU}{VLU}$

NO

62    $ULU \leq VS(7)$?   YES $TLU = \dfrac{Q(7) - [VS(7) - ULU] \cdot T(4)}{VLU}$

NO

63    $ULU \leq VS(8)$?   YES $TLU = \dfrac{Q(8) - [VS(8) - ULU] \cdot TLL}{VLU}$

NO

64   EXCESSIVE FLOW

$$Q(8) = Q(8) - VLU \cdot TLU$$
$$Q(9) = Q(9) - VLU \cdot TLU$$

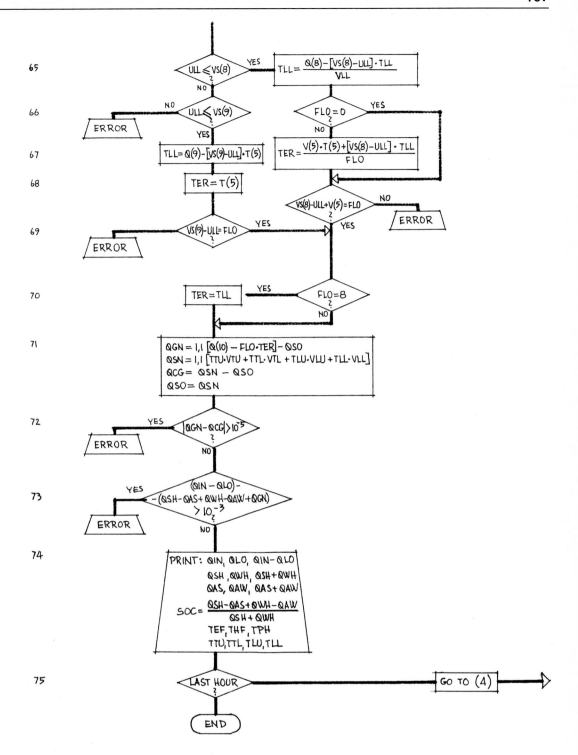

65 ULL ≤ VS(8)? YES $TLL = \dfrac{Q(8) - [VS(8) - ULL] \cdot TLL}{VLL}$

NO

66 ULL ≤ VS(9)? NO → ERROR YES

FLO = 0? YES

NO

67 $TLL = Q(9) - [VS(9) - ULL] \cdot T(5)$ $TER = \dfrac{V(5) \cdot T(5) + [VS(8) - ULL] \cdot TLL}{FLO}$

68 $TER = T(5)$

$VS(8) - ULL + V(5) = FLO$? NO → ERROR

YES

69 $VS(9) - ULL = FLO$? YES → ERROR

70 $TER = TLL$ YES FLO = 8?

NO

71
$QGN = 1,1 \, [Q(10) - FLO \cdot TER] - QSO$
$QSN = 1,1 \, [TTU \cdot VTU + TTL \cdot VTL + TLU \cdot VLU + TLL \cdot VLL]$
$QCG = QSN - QSO$
$QSO = QSN$

72 $|QGN - QCG| > 10^{-5}$? YES → ERROR

NO

73 $(QIN - QLO) - (QSH - QAS + QWH - QAW + QGN) > 10^{-3}$? YES → ERROR

NO

74
PRINT: QIN, QLO, QIN−QLO
QSH, QWH, QSH+QWH
QAS, QAW, QAS+QAW
$SOC = \dfrac{QSH - QAS + QWH - QAW}{QSH + QWH}$
TEF, THF, TPH
TTU, TTL, TLU, TLL

75 LAST HOUR? → GO TO (4)

END

168

Symbols used

Constants:	KSH	coeff. of heat transfer in space heating fan-coil
	KWH	coeff. of heat transfer in water pre-heater coil

Temperatures (°C):

TST	starting temperature
TEO	outdoor air temperature
TMP	mean plate temperature
TPP	same in 'previous' hour
TEF	flow temperature (from plate)
TER	return temperature (to plate)
TTU	top, upper ⎫
TTL	top, lower ⎬ layers in storage
TLU	lower upper ⎪
TLL	lower lower ⎭
T(1)–T(5)	dummy storage temperatures
TD1–TD5	temperature difference settings
THF	flow temperature in heating circuit
TSN	space heating 'normal' return temperature
TSR	space heating return temperature
TPH	temperature in pre-heating cylinder
TWR	water heating return temperature
TRM	room temperature

Flow quantities (l/h):

FLO	in collection circuit
FTO	total flow into heating circuit
FWH	in water heating branch
FSH	in space heating branch
FSR	in space heating return
FCW	'consumed' water (in H/W system)
FPH	set value for FWH
FL1 ⎫	low and high settings for FLO
FL2 ⎭	

Volumes (l):

VTU	top, upper ⎫
VTL	top, lower ⎬ storage layers
VLU	lower, upper ⎪
VLL	lower, lower ⎭
V(1)–V(5)	⎫
VS(1)–VS(9)	⎪
UTU	⎬ dummy variables for stratification in storage
UTL	⎪
ULU	⎪
ULL	⎭

Time:

HON	hour 'on' ⎫ time settings for boiler
HOF	hour 'off' ⎭
TIM	fraction of hour worked for space heating pump

Heat flow quantities (W):

QIN	radiation absorbed
QLO	heat loss from collector
QSM	space heating max. output
QSH	space heating output
QGN	heat gain or loss by storage
Q(1)–Q(10)	dummy variables for stratification
QSO	heat content of storage (Wh)
QSN	'new' value for QSO (Wh)
HCP	heat capacity of collector (Wh/degC)
HLR	heat loss rate of house (W/degC)
QAS	auxiliary to space heating
QAW	auxiliary to water heating
QWH	total water heating

Areas (m²):	ARP	collector plate
	AWH	water pre-heating coil
	ASH	space heating coil

Others:	ATS	attenuation coeff. in space heating coil
	ATK	attenuation coeff. in water heating coil
	SOC	solar contribution factor

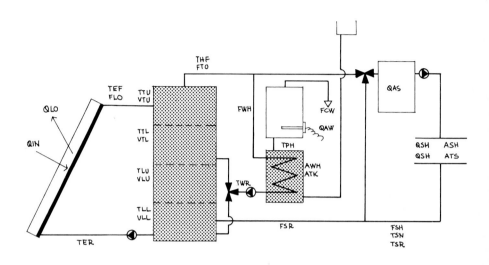

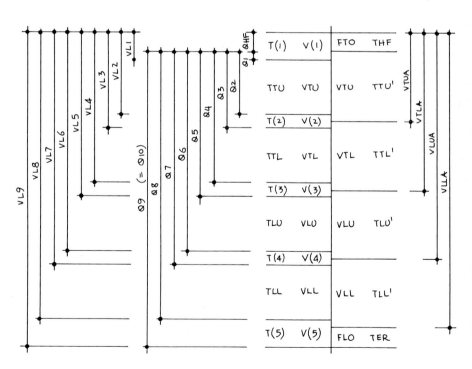

Dummy variables for temperature
stratification in the storage tank.

SUBJECT INDEX

The author wishes to express his gratitude to Miss Maria Kovàcs for preparing this index and for revising it for the second edition